A Biographical Dictionary of American Civil Engineers

Volume II

Committee on the History and Heritage of American Civil Engineering
of the American Society of Civil Engineers

Francis E. Griggs, Jr., Editor
Karen O'Connor, Research Assistant

 Published by the
American Society of Civil Engineers
345 East 47th Street
New York, New York 10017-2398

ABSTRACT

This publication *A Biographical Dictionary of American Civil Engineers—Volume II* supplements the first volume of this biographical dictionary with over 200 additional entries. While a civil engineer needed to be born prior to the Civil War in order to be considered for inclusion in volume I, for this second volume the basic criteria is that the engineer be born before 1900. The entire selection process is guided by the definition of a "civil engineer" as one who engages in the planning, design, and supervision of construction of civil works. This includes those who provide the surveys for these civil works. Those selected for inclusion were chosen based on the frequent references to them and their work in publications on engineering history. Biographical files on American civil engineers located in the Division of Engineering and Industry at the Smithsonian Institution's National Museum of American History were also consulted. Civil engineers, educators, and historians will find this volume to be as useful as the first volume. In addition, it should encourage greater interest and more scholarly investigation into the history of civil engineering.

Library of Congress Cataloging-in-Publication Data

A Biographical dictionary of American civil engineers.
  (ASCE historical publication)
  Vol. 2: Francis E. Griggs, Jr., editor; Karen O'Connor, research assistant.
  1. Civil engineers—United States—Biography—Dictionaries. I. Griggs, Francis E. II. O'Connor, Karen. III. American Society of Civil Engineers. Committee on History and Heritage of American Civil Engineers. IV. Series: American Society of Civil Engineers. ASCE historical publication.
TA139.A53   624'.092'2   [B]   72-194203
ISBN 0-87262-822-1

The Society is not responsible for any statements made or opinions expressed in its publications.

Authorization to photocopy material for internal or personal use under circumstances not falling within the fair use provisions of the Copyright Act is granted by ASCE to libraries and other users registered with the Copyright Clearance Center (CCC) Transactional Reporting Service, provided that the base fee of $1.00 per article plus $.15 per page is paid directly to CCC, 27 Congress Street, Salem, MA 01970. The identification for ASCE Books is 0-87262/88. $1 + .15. Requests for special permission or bulk copying should be addressed to Reprints/Permissions Department.

Copyright©1991 by the American Society of Civil Engineers, All Rights Reserved.
Library of Congress Catalog Card No: 72-194203
ISBN 0-87262-822-1
Manufactured in the United States of America.

CONTENTS

PICTURE CREDITS..v

FOREWORD..vii

INTRODUCTION..ix

PREPARATION OF VOLUME III..xi

GUIDE TO REFERENCES..xiii

BIOGRAPHIES..1

INDEX OF BIOGRAPHIES...130

PICTURE CREDITS

Page

10	ASCE Files
15	ASCE Files
27	ASCE Files
28	The Smithsonian Institution
40	Office of History, US Army Corps of Engineers
46	Clyde Gessel
51	The Smithsonian Institution
64	ASCE Files
68	ASCE Files
76	Office of History, US Army Corps of Engineers
84	ASCE Files
87	Parsons, Brinckerhoff, Quade & Douglas, Inc., New York, NY
96	The Smithsonian Institution
101	ASCE Files
105	ASCE Files
108	Eddie B. Smith, Lexington, KY
114	The Smithsonian Institution
118	The Smithsonian Institution

FOREWORD

The American Society of Civil Engineers' Committee on the History and Heritage of American Civil Engineering is again pleased and proud to write this foreword for <u>A Biographical Dictionary of American Civil Engineers-Volume II</u>. This volume was funded by a grant from ASCE and was prepared at Merrimack College under the editorship of Dr. Francis E. Griggs, a member of the committee, with research assistance from Karen O'Connor, a recent graduate in history from Merrimack. The list of possible entries was assembled from names compiled in Volume I and from recommendations by members of the committee and of ASCE. As in the preparation of the first volume, many apparent conflicts of fact had to be resolved and the technical ramifications of each engineer's work had to be put into perspective in relatively few words. We have made every attempt to eliminate errors and resolve factual conflicts. We believe that Dr. Griggs and Ms. O'Connor have done an outstanding job.

We hope the Volume II will be as useful to the Society as was Volume I when it was published in 1972, and that it encourages greater interest and more scholarly investigation into the field of civil engineering history.

Committee on the History and Heritage of
American Civil Engineering for 1989-1990

David P. Billington
J. E. Colcord
Clyde D. Gessel
Francis E. Griggs
Joseph E. Minor
Howard H. Newlon, Jr.
Alan L. Prasuhn
Neal FitzSimons, Chairman

INTRODUCTION

This volume of the <u>Biographical Dictionary</u> contains 218 entries, supplementing the 170 entries contained in Volume I. To be considered for inclusion in Volume II, the engineer must have been born prior to 1900, whereas in Volume I the engineer must have been born before the Civil War. Those selected for inclusion in this volume were chosen by the editors of Volume I and members of ASCE's Committee on the History and Heritage of American Civil Engineering on the basis of frequent references to them and their work in publications on engineering history. Also consulted were the biographical files on American civil engineers located in the Division of Engineering and Industry at the Smithsonian Institution's National Museum of American History (NMAH) in Washington, D.C.

The selection process for persons on whom entries were planned was complicated by the fact that many engineers who practiced in the nineteenth century frequently did not go by the title of "Civil Engineer." Many individuals also worked in other engineering fields and often in the management of the railroads. We continued the precedent set in Volume I, and upon occasion were somewhat arbitrary in deciding whether or not to call the individual a civil engineer.

In this volume of the <u>Biographical Dictionary</u>, as in Volume I, we defined a "civil engineer" as one who engages in the planning, design and supervision of construction of civil works. These works are the facilities associated with transportation and communication, the utilization and protection of resources, and urbanization. Typical structures and systems included are canals, railroads, bridges, tunnels, dams, waterworks, sewage treatment plants, levees, irrigation systems, terminals, and large buildings. Also associated with the works are a variety of surveys which normally are accomplished by civil engineers. It also seemed desirable to include a few selected individuals who as teachers, inventors, researchers, or scientists directly contributed to the development of civil engineering.

Some of the engineers found in this dictionary have entries in other, more general biographical works, although greater detail of their professional engineering careers will be found in this work.

In the preparation of entries, several conventions were adopted for the sake of brevity that would still maintain the same level of accuracy of the information included in the entry.

1. A number of professional and military designations (e.g. Asst. Engr.) were used and every effort was made to place in perspective the role of the engineer on any project mentioned.

2. It is often the case that the date is known when an engineer began his association with an institution or project, but not when he left. For this reason, entries will occasionally read "1868- ", rather than "ca. 1868."

3. When the years of attendance at an educational institution are given, it can be assumed that graduation occurred in the last year unless "did not graduate" is noted in the entry.

4. After the full name of an enterprise or project is provided in an entry, an obvious abbreviation will be used thereafter. For example, "Delaware & Hudson Canal" will be referred to subsequently as "D&H". All railroads will be abbreviated "RR", even though their original name might have been "railway".

5. When a particular date is in doubt, the earliest possibility is given, followed by any others in parenthesis.

6. All graduates of the United States Military Academy mentioned in this volume are also in Cullum (see Guide to References, p. viii)

7. When the length of a structure is given, the reader may assume that this is the total length unless otherwise indicated.

Readers who have additional information about engineers described in this volume or have memorabilia relating to civil engineers are encouraged to contact ASCE's Coordinator of Historical Activities, 1015 15th St., N.W., Suite 600, Washington, DC 20005 or the Division of Engineering and Industry at NMAH, The Smithsonian Institution, Washington, DC 20560.

PREPARATION OF VOLUME III

The Committee on The History and Heritage of American Civil Engineering is planning a third volume of the <u>Biographical Dictionary</u>. The scope of the work is to be international and all civil engineers born prior to 1925 will be considered for inclusion. Readers are encouraged to suggest engineers to be included in the next volume by submitting a brief biography of their suggestions in the following form:

1. Name, date and location of of birth, and date and location of death.

2. Father's name, mother's name.

3. Wife's name.

4. Number and/or names of children.

5. Education background.

6. Chronological description of professional experience.

7. Membership in professional societies.

8. Awards, honors and/or honorary degrees.

9. References to engineer's published works.

10. References to works that describe his or her career, for example, other biographical dictionaries.

Please submit all nominees to the Coordinator of Historical Activities, American Society of Civil Engineers, 1015 15th St., N.W., Suite 600, Washington, DC 20005.

GUIDE TO REFERENCES

ASM - *America's Successful Men of Affairs*, by Henry Hall, 1895-96.
BDNA - *The Twentieth Century Biographical Dictionary of Notable Americans*, edited by Johnson Rossiter. Boston: Boston Biographical Society, 1904.
BOS - Bosley's *History of King County*, vol. 2. Chicago, 1929.
CAB - *Appleton's Encyclopedia*.
CCI - *American Building*, vol. 1, by Carl Condit.
CE1 - *Civil Engineering* (ASCE), Jan. 1941, p. 60.
CE2 - *Civil Engineering* (ASCE), Mar. 1949, p. 193.
Cullum - *Biographical Register of the Officers and Graduates of the U.S. Military Academy at Wet Point, New York*, 3rd edition, by George W. Cullum, 1891.
DAB - *Dictionary of American Biography*.
DR - *Dictionary of American Biography*, by Francis Drake.
EOB - *Encyclopedia of Biography*.
GEPT - *Great Engineers and Pioneers in Technology*, vol. 1, by Roland Turner and Steven Gould. New York: St. Martin's Press, 1981.
Lamb - *Lamb's Dictionary of American Biography*, edited by Brown.
MP - *Public Works in Seattle: A Narrative History of the Engineering Department*, by Myra Phelps. Seattle: City of Seattle, 1976.
PASCE - *Proceedings of the American Society of Civil Engineers* (memoirs)
RHT - *That Man Thompson*, by R. H. Thompson, edited by Grant Redford. Seattle: Univ. of Washington Press, 1950.
RPI - *Biographical Record of the Officers and Graduates fo the Rensselaer Polytechnic Institute, 1824-86*, edited by Henry Nason. Troy, NY: William A. Young, 1887.
TASCE - *Transactions of the American Society of Civil Engineers* (memoirs).
WAB - *National Cyclopedia of American Biography*, edited by White.
WWE - *Who's Who in Engineering, 1922-37*.
WWW - *Who Was Who in America*.

Aldrich - "The New Technology," Aldrich and Seeler.
Bray - Private files of Oscar Bray.
Gessel - Private files of Clyde Gessel.
Kemp - "James Finley and the Modern Suspension Bridge," Emory Kemp.
Morris Knowles - Private files of Morris Knowles & Assoc.
Poirier - Clarisse Poirier
Schnitter - "The Evolution of the Arch Dam," by N. J. Schnitter.
Wolman - Vita of Abel Wolman.
YBP - Private Files of Yamin, Parsons, Brinckerhoff.

ABRAMS, DUFF ANDREW; b. in IL. Grad. Univ. of Illinois, 1905. Engr. Experimental Station, 1905-14. Dir. of Research, Portland Cement Assn., now Dir. of Research, Intl. Cement Corp. Made tests in all phases of concrete and reinforced concrete including notable tests on bond. Developed idea of "water-cement ratio" in proportioning of concrete and invented, with E. H. Harnder, the "colorimetric test" for impurities in sand.

ALDEN, JOHN FERRIS; b. Cohoes, NY, March 19, 1852; d. Rochester, NY, Feb. 27, 1917; f. Sidney; m. Harriet Webster; w. Mary E. Bogue; c. 2 sons and 3 daughters. Educated in private schools in Albany, NY, entered RPI, Sept. 1868. Grad. June 1872 with degree of civ. engr. Asst. engr. on 2 bridges over Hudson River at Albany built by NY Central & Hudson River RR and Boston & Albany RR, or under a company known as Hudson River Bridge Co. Remained as asst., designed bridge work until Jan. 1, 1874. Asst. engr. at Leighton Bridge and Iron Works, Rochester, NY. Chief engr. in 1877, partner in 1878. Leased Leighton Bridge and Iron Works, continued business in partnership with Moritz Lassig, July 1, 1881; in Rochester under name of Alden & Lassig, bridge builders and contractors, and in Chicago under name of Lassig & Alden. Partnership dissolved in Jan. 1886, Alden purchased business interests in Rochester, and Lassig in Chicago. Reorganized under name of Rochester Bridge and Iron Works. Sold in 1901 to American Bridge Co. Bridges built during active connection with Rochester Bridge and Iron Works include: elevated RR in NYC, bridges for Delaware and Hudson RR, Chicago, Milwaukee and St. Paul RR and Buffalo, Rochester and Pittsburgh RR, bridge over the Columbia River at Pasco, WA, viaducts at Los Angeles, CA, upper suspension bridge at Niagara Falls, and Driving Park Ave. Bridge at Rochester, NY. Also const. tower and elevator in House of Parliament at Ottawa, Ontario, Canada; buildings in NY, Chicago and other cities; furnished iron and steel work for buildings of Columbian Exposition at Chicago. Pres., Locke Insulator Manufacturing Co. and dir., Genesee Valley Trust Co. Mem: ASCE, Rensselaer Soc. of Civ. Engrs., Rochester Chamber of Commerce, Alden Kindred of Am., Sons of the Mayflower, Genesee Valley Club, Rochester Club, Friendly Home (vice pres., Bd. of Trustees). Refs: TASCE, DAB, RPI Bio.

ALLEN, CALVIN FRANCIS; b. Roxbury, MA, July 10, 1851; d. June 6, 1948; f. Calvin; m. Ann Priscilla (Watson); w. Caroline Elizabeth Hadley; c. Mildred, Margaret, Frances. Grad. Roxbury Latin, 1868. Grad. MIT in 1872 with degree of civ. engr. Asst. engr. water works and sewers, Providence, RI and Newton, MA, 1872-78. Asst. engr., Atchison, Topeka and Santa Fe RR, 1878-85, except one year chief engr., water works at Las Vegas, NM. Admitted to practice in courts of New Mexico, 1885. Attorney for Atchison Topeka and Santa Fe at Socorro, NM, 1886. Asst. prof., RR engr. at MIT, 1887. Made assoc. prof., served as secy. for alumni assn. 1887. Mem. school committee, Sharon, MA, 1892-95, chair, 1893-95. Retired under Carnegie Foundation, 1916, prof. emeritus. Household fuel economy agent for MA in US Fuel Administration, Sept. 1-Dec. 31, 1918. Mem. Advisory Zoning Commission for Boston, 1922-23. Engr.D. Northeastern Univ., 1938. Mem: ASCE, Boston Soc. Civ. Engrs. (pres. and hon. mem.), Mass Highway Assn. (pres.), New England RR Club (pres.), Am. Railway Assn. (life mem.), Soc. for the Promotion of Engr. Education (secy., pres.), Technology Club (secy.), Committee on Public Utilities, Boston Chamber of Commerce. Publ: *Railroad Curves and Earthwork* (1889), *Tables for* *the Computation of Earthwork* (1893), *Field and Office Tables* (1903), *Business Law for Engineers* (1917). Refs: Lamb, WWW, BDNA, WWE.

ALLEN, KENNETH; b. New Bedford, MA, April 6, 1857; d. Sept. 7, 1930; f. Edward Augustus; m. Eugenia Sophia Teulon; w. Rose Whitmore Switzer; c. Edward Switzer, Harold Ames, Russell, Francis Eleanor. Prep school of Univ. of Wisconsin, Madison, 1872-73. Grad. West Newton, M.A., English and Classical School, 1874. Entered RPI, June 1874, grad. with degree in civ. engr., 1879. Employed in Boston Water Works, Sudbury River, Framingham, MA, July 1875-Aug. 1877. Employed in Essex County, Lawrence, MA, Aug. 1879-Sept. 1879. Engaged in railway engr. work in Kansas City, St. Joseph. & Council Bluffs RR, St. Joseph, MO, Sept. 1879-Feb. 1881. Worked for Missouri Pacific RR, Mineola, TX, Feb. 1881-May 1881. Building Dept., Atchison, Topeka & Santa Fe RR, Topeka, KS, May 1881-May 1882. Pittsburgh, McKeesport & Youghiogheny RR, Pittsburgh, PA, May 1882-May 1883. Asst. engr., Philadelphia Water Dept. of Surveys, Philadelphia, PA, May 1883-86. In charge Division 2, Nodaway Valley RR; Division 3, Tarko Valley RR (branches of Kansas City, St. Joseph and Council Bluffs RR); Division 4, Greenville, Mineola extension of Missouri Pacific RR. Made designs for masonry for Pittsburgh, McKeesport & Yioughioneny RR. Breithaupt & Allen, const. engr. offices in Kansas City, MO. Supt. of const. for Kansas City, MO 1886-90. Engr. in charge of topographical survey of Connellsville coke region for H.C. Frick Coke Co., Scottdale, PA, 1890-93. Asst. engr. in charge public works, Yonkers, NY, 1893-95. Asst. engr. on sewerage system of Williamsbridge, NY. In charge of studies for sewerage sstem and sewage disposal of Baltimore, MD, 1895-1900. Consulting practice with N.S. Hill and A.M. Quick under name of Hill, Quick and Allen of Baltimore and NYC, reporting on sundry water supply and sewerage projects, 1900-1902. Engr. and supt., Water Dept. of Atlantic City, NJ, 1902-1906. Engr. in charge of Low-Level Division of the sewerage system in Baltimore, 1906-1908. Principal asst. engr., Baltimore Sewerage Commission, New York, 1908-14. Studies of NY harbor and investigations of sewers and sewage disposal in metropolitan district until 1913. During 1913, inspected sewage systems and sewage disposal works abroad including England, Scotland, Holland, France and Germany. Engr., Bureau of Sewers Planning, NY, 1915-16. Engr. sewage disposal and then sanitary engr., Bd. of Estimate and Apportionment of NY, 1916-30. During leave of absence during World War I, served as district engr., US Housing Corp., Washington, DC. Sanitary engr., Sanitation Commission, NY 1930. Mem: ASCE, Sigma Xi, Am. Soc. for Municipal Improvements, Am. Public Health Assn. Municipal Engrs. of the City of NY, NY State Sewage Works Assn. (pres.), Engrs. Soc. Western PA, Rensselaer Soc. of Engrs., Philadelphia Engrs. Club; Unitarian Laymen's League, Tramp and Trail Club of NY, Westchester Trails Assn., Univ. Club of White Plains, NY. Publ: *Sewage Sludge* (1912), "Clarification of Sewage by Fine Screening" (1915), "Transmission of Typhoid by Sewage-Polluted Oysters," "The Pressing of Sewage Sludge" and other technical papers. Refs: RPI Bio, TASCE, WWW, WWE.

AMMANN, OTHMAR HERMANN; b. Fuerthalen, Switzerland, March 26, 1879; d. NYC, Sept. 26, 1965; f. Emmanuel Christain Ammann; m. Emilia Rose Labharot; w. (1st) Lilly Selma Wehrli (died 1932); w. (2nd) Klary Noetzli; c. Werner, George Andrew, and Margot. Grad. Swiss Fed. Polytechnic Inst., Civ.

Engr. 1902; employed as structural draftsman, Wartmann and Valette, Brugg, Switzerland (1901-1903) participating in topographical surveying for mountain RR from Montreaux to the Bernese Oberland. Asst. engr., Buchheim and Heister, Frankfurt, Germany (1903-1904) employed in the design of reinforced concrete structures. On the advice of teacher, Prof. Karl Emil Hilgard, emigrated to US in 1904 to work on RR bridges. Engr. asst. to Joseph Mayer, Union Bridge Co., NY, working on approx. 30 steel RR bridges. Asst. engr. PA Steel Co., Steelton, PA, working on steel bridges and bldgs. (1904-1909), including Lindenthal's Queensboro Bridge, NY. In 1906, designed steel bridges for Ralph Modjeski in Chicago. Principal asst. engr. for C. C. Schneider and Frederick C. Kunz, consulting engrs., Philadelphia, PA (1909-12) including investigations into St. Lawrence cantilever bridge failure at Quebec (1908) and replacement design, as well as St. John River arch bridge, New Brunswick, Canada. Principal asst. engr. to Gustav Lindenthal const. engrs., NY (1912-23), as well as asst. chief engr. of the NY Connecting RR and the North River Bridge Co.; supervised design and const. of the Hell Gate Bridge, NY; the Sciotoville Bridge over the Ohio River, OH; studies for proposed Hudson River crossing at 57th St., NYC. Est. self as consulting engr. (1923-25) developed preliminary designs for George Washington Bridge. Appt. Chief Engr. of bridges; Chief Engr., then Dir. of Engr., Port of NY Authority (1925-39). Designed and supervised const. of the George Washington Bridge across the Hudson River (3500-ft. record suspended span, 1931); Outerbridge Crossing and the Goethals Bridge across the Arthur Kill (1928); Bayonne Bridge (1652-ft. record arch span, 1931); the Lincoln Tunnel connecting NY and NJ (8,216-ft. length, 1937). Bd. of engrs. for design and const. of the Golden Gate suspension bridge, San Francisco, CA (4200-ft. record span, 1929-37). Chief engr., Triborough Bridge and Tunnel Authority, NYC (1934-39) working closely with Chair Robert Moses on the planning and const. of the Triborough Bridge (1936) and the Bronx-Whitestone Bridge (1939). Consulting engr., 1939-66. Bd. of engrs. investigating Tacoma Narrows Bridge failure for Federal Works Admin. (1939). In partnership with landscape architect C.C. Combs, prepared and supervised const. of the NY-NJ Palisades Interstate Pkwy., as well as several hundred miles and 60 bridges for several other NY Metro area highways, including key sections of the NJ Turnpike. After 1946, in partnership with Charles S. Whitney of Milwaukee, WI, est. Ammann & Whitney, Consulting Engrs., designed Verrazano-Narrows Bridge (record 4260-ft. suspended span, 1964), Delaware Memorial suspension bridge across the Delaware River at Wilmington, DE (1951); collaborated with Modjeski & Masters on Walt Whitman Bridge (1957); TransWorld Airways Intl. Terminal, JFK Intl. Airport; Pittsburgh Arena (415-ft. diam. retractable domed roof, 1963); as well as a 2nd deck for the George Washington Bridge, and numerous bomb-resistant structures for the US military, auditoriums, airports and hangars in the US, Europe, Africa, and Asia. Publ: The Hell Gate Arch Bridge Over the East River in New York City (1918); The Problem of Bridging the Hudson River in New York (1924); The Bridge Across the Kill van Kull at New York (1930); George Washington Bridge - General Conception and Development of Design Transactions (1933); Der Lincoln Tunnel unter des Hudson in New York (1938); Planning and Design of the Bronx-Whitestone Bridge (1939); The Failure of the Tacoma Narrows Bridge (1941); Bridges of New York (1945); Brooklyn Bridge after 60 Years (1947). Awards: ASCE's Thomas Fitch Rowland Prize in 1918 for Hell Gate Bridge; Ernest E. Howard Award in 1960 "for design and construction of outstanding bridges of record dimensions and outstanding beauty;" Fellow, Am. Assn. for Adv. of Science (1931); Distinguished Serv. Award, Am. Soc. for Metals Distinguished Service Award in 1948 for contributions to the use of high-strength steels for long-span bridges; 1st prize in Annual Award Competition of the Am. Inst. of Steel Const. in 1931 for the Bayonne Bridge, in 1936 for the Triborough Bridge, in 1937 for the Golden Gate Bridge, in 1938 for the Little Hell Gate Bridge, and in 1939 for the Bronx-Whitestone Bridge; Engr. Award for Outstanding Achievements, NJ Soc. of Prof. Engr.; Civ. Engr. Award (1958); Award of Merit (1961); Allied Professions Gold Medal (1962), NY Chapt. Am. Inst. of Architects, for aesthetics in bridge design; and the National Medal of Science from Pres. Lyndon Johnson in 1964, the 1st presentation to a civ. engr. Hon. degrees: D.Sc., Swiss Federal Polytechnical Inst. (1930); D.Engr., New York Univ. (1931); D.Sc., Yale Univ. (1932). D.Engr., Pennsylvania Military Academy (1937); D.Sc., Columbia Univ. (1941); and D.Sc., Brooklyn Polytechnical Inst. (1956). Mem: ASCE (hon. mem., 1953 and dir., 1934-36), Am. Acad. of Arts and Sciences (fellow), Swiss Soc. of Engr. and Architects (hon. mem.), Tau Beta Pi, Chi Epsilon, Am. Inst. of Consulting Engr., Inst. of Civ. Engr. of Great Britain, Am. Soc. for Testing Materials, Am. Soc. for Testing Materials, Am. Railway Engr. Assn.; Natl. Soc. of Prof. Engr.

ANDERSON, JOHN FRANCIS; b. Jemshog, Sweden, Dec. 30, 1848; d. Jan. 23, 1927; f. Anders Anderson Thore; m. Ingar Svenson; w. Cecilia Anderson. Self-educated, came to US as a sailor, 1869. Assisted in const. of bridge across Missouri River for Union Pacific RR, 1870. Became foreman-supt. for bridge work, building piers and foundations. Built river piers for South Street Bridge, Philadelphia, 1872. Foundations for Iron Mountain RR Bridge over Arkansas River at Little Rock, 1873. Employed by govt. of Venezuela, 1875. Engaged in bridge work in England, 1876-79. Supt., Hudson River Tunnel between NY and Jersey City, 1879-82. Built bridge over Atchafalaya River, LA for Texas Pacific RR, 1882-84. Head of firm Anderson & Barr, 1884-95. Const. foundations for bridges including those over Arkansas River at Little Rock, across Ohio River at Cairo, Merchant's Bridge over the Mississippi at St. Louis and across St. John's River at Jacksonville, FL. Other const. included a 12-ft. drainage tunnel one mile long, a 15-ft. drainage tunnel 2 miles long in Brooklyn, bridge foundations and a lighthouse in Delaware Bay. Const. deepest bridge foundations in the world for the Hawkesbury Bridge in Australia. Inventor and owner of patents for aerial bridges and the pilot system of tunneling. Decorated Merit of Military, Spain, Knight Cmdr. Order of Vasa, Sweden. Refs: WWW, WWE.

BAKER, GEORGE TITUS; b. Iowa Co., IA, July 9, 1857; d. Dec. 13, 1940; f. Albert Watson; m. Freelove Malicent Kenyon; w. Clara L. Poole; c. Ethel, Georgia E., Sue A. Preparatory education at Hall's School for Boys, Ellington, CT and McClain's Academy, Iowa City, IA. Attended State Univ. of Iowa and Cornell Univ., 1895-97. Division engr., Connecticut, Rhode Island and Pennsylvania RR, 1879-85. Locating and const. engr., Atchison Topeka and Santa Fe RR, 1886-88. Chief engr., Soo and Southwestern RR, 1888-89. Chief engr., Muscatine High Bridge, Clinton High Bridge. Const. engr., Winona High Bridge, 1889-91. Mgr. and chief engr., Edwards & Walsh Const. Co. and Tri-City Const. Co, 1892-1912. Vice pres., Clinton St. RR Co., Choctaw Lumber Co. Mem. Iowa House of Representatives, 1896-98, mayor of Davenport, IA, 1898-

1900, IA State Bd. of Education, 1909. Appt. mem., RR Emergency Bd. Refs: WWW.

BAKER, WILLIAM EDGAR; b. Springfield, MA, Oct. 18, 1856; d. NYC, Nov. 7, 1921; f. Henry Martyn; m. Susan Virginia (Barnes); w. Harriet Estelle Griffin; c. William Edgar, Phoebe, Eugenia G., Charlotte S., and Alan Griffin. Grad. Lafayette College, 1877, degree of civ. engr. Transitman and engr. in charge of work, St. Paul & Pacific RR, 1877-81. Transitman on surveys for line through the Rocky Mountains, Canadian Pacific RR, 1881-83. Resident engr. in charge of bridges and buildings of the Intl. & Great Northern RR, 1883-84. Resident engr. and supt. of the Missouri Pacific RR Co., 1884-89. Supt. of Electric Service of the Thompson Houston Electric Co. (now General Electric Co.) in Boston, MA, in charge of the installation of the electric equipment of the West End Street RR, 1889-92. Chief engr. and gen. mgr., Columbia Intermural RR, 1892-94. Gen. mgr. and chief engr., Metropolitan West Side Elevated RR, Chicago, 1894-99. Chief electrical engr. in charge of const. and installation of an electrical system of the Manhattan RR, NY, 1899-1902. Senior member, W.E. Baker & Co., const. engrs., 1902. Designed and supervised the const. of the electrical equipment of the Calumet and Hecla Copper Mines. Const. the first bascule bridge over the Chicago River and was consulted for const. of the London Underground RRs. Mem: ASCE; Am. Soc. of Mech. Engrs.; assoc. mem., Am. Inst. of Electrical Engrs.; Engrs. Club of NYC; Trustee, Lafayette College. Refs: WWW, TASCE, WAB, WWE.

BARNARD, JOHN FISKE; b. Worchester, MA, April 23, 1829; d. Los Angeles, CA, Feb. 6, 1910; f. John; m. Sarah R. (Bigelon); w. Gertrude Harvey; w. 2nd, Julia Keefer: c. 11. Attended the common schools, also Bridgewater Normal School. Entered RPI, Nov. 7, 1849. Began professional work in Nov. 1850 on the St. Lawrence and Atlantic RR. Asst. engr. on const. and resident engr. on that road for seven years. For five years was supt. of the Montreal and Champlain RR. Supt. of Buffalo and Lake Huron Division of Grand Trunk RR, 1864-66. Chief engr. of the Grand Trunk RR 1866-69. Supt. and chief engr. of the Missouri Valley RR, 1869-70. Chief engr. of the Kansas City, St. Joseph & Council Bluffs RR, July 1, 1870- June 1871. Chief engr. and gen. supt. of the St. Joseph and Denver City RR, June 1871-April, 1872. Gen. supt. of St. Joseph and Denver City RR, Aug 1872- . Appt. gen. supt. of the St. Joseph and Des Moines RR, July 1881- . Appt. gen. mgr. of Hannibal and St. Joseph RR, April 1, 1884. Designed and built 50 miles of St. Joseph and DC RR and four branches of Kansas City, St. Joseph and Council Bluffs RR. Secy. and treas. of St. Joseph Union Depot Co., pres. of Atchinson Union Depot and RR Co., pres. of Union Stockyard Co., St. Joseph, MO. Appt. pres. and gen. mgr., Ohio & Mississippi RR, 1886-92. Engaged on several reports of projected RRs and appraisals of industrial and RR properties, 1892-93. Receiver of the Omaha and St. Louis RR and pres. of the Alton Bridge Co. and receiver of the St. Clair-Madison and St. Louis Belt Line, 1893-98. Mem: ASCE, Am. Geographical Soc. Refs: RPI Bio, PASCE.

BARNARD, JOHN GROSS: b. Sheffield, MA, May 19, 1815; d. Detroit, MI, May 14, 1882; f. Robert Foster; m. Augusta (Porter); w. Jane Elizabeth Brand; m. 2nd Anna E. (Hall) Boyd. Received appt. to US Military Academy, entered July 1, 1829. Grad. 2nd in class of 43, 1833. Assigned as brevet 2nd lt.

to U.S. Army Corps of Engrs., Newport, RI. Assigned to coast defenses, Pensacola and New Orleans, 1833. A.M., Univ. of Alabama, 1838. Attained capt., 1848. Supervising engr., fortifications of several ports including NY, Portland and Mobile during 1840's. Supervised fortifications at base of Tampico and surveyed battlefields around Mexico City during Mexican War, brevetted maj. for services, May 30, 1848. Chief engr. for RR survey across isthmus of Panama for Tehuantepec RR Co. of New Orleans, 1850. Report was first topographical account of the isthmus. Surveyed mouth of Mississippi River, 1852. Instr. of practical engr. at Military Academy. Supt., Military Academy, 1855-56. Placed in charge of fortifications of New York Harbor. Promoted to maj. of engrs., Dec. 13, 1858. Chief engr. of Dept. of Washington, D.C., following beginning of Civil War, April-July 1861. Chief engr. to Gen. McDowell in the First Bull Run campaign. Commissioned brig. gen. of Volunteers, 1861, chief engr. of the Army of the Potomac, 1862. Chief engr. for Gen. McClellan during Peninsular campaign, 1862. LL.D. Yale College, 1864. Brevetted maj. gen., 1864. Chief engr. all field armies on Gen. Grant's staff, 1864. Nominated as Chief of Corps of Engrs. in 1864, at own request name was withdrawn. Promoted col. of Corps of Engrs., Dec. 28, 1865. Chief advisor, Bd. on Improvements of Mouth of Mississippi River, 1871. Pres. of Permanent Bd. of Engrs. for Fortifications and River and Harbor Improvements. One of original incorporators of the National Academy of Sciences. Retired 1881. Publ: *Survey of the Isthmus of Tehuantopec* (1852), *Phenomena of the Gyroscope Analytically Examined* (1858), *Outlets and Levees of the Mississippi River* (1859), *Dangers and Defenses of New York* (1859, "Memoir on National Defenses" (1860), *Notes on Sea Coast Defense* (1861), *The Confederate States of America and the Battle of Bull Run* (1862), *Reports of the Engr. and Artillery Operations of the Army of the Potomac* (1863), *The Peninsular Campaign and its Antecedents as Developed by the Report of Maj. Gen. George B. McClellan and Other Published Documents* (1864), *Eulogy on General Totten* (1866), *Report on the Defenses of Washington* (1871), *Report on the Fabrication of Iron For Defensive Purposes and Its Uses in Modern Fortification Especially in Coast Defense* (1871-72), "The North Sea Canal of Holland and Improvement of Navigation from Rotterdam to the Sea," *Problems of Rotary Motion presented by Gyroscope, the Precession of the Equinoxes and the Pendulum* (1873), *On the Internal Structure of the Earth Considered as Affecting the Phenomena of Precession and Nutation* (1877), *An Alleged Error in Laplace's Theory of the Tides* (1877), *Some Remarks on the Use and Interpretation of Particular Integrals Which "Satisfy" General Differential Equations Expressive of Dynamic Problems in Cases where General Integration is Impossible* (1877). Refs: Lamb, DAB, WWW, CAB, WAB.

BARNES, OLIVER WELDON; b. Berlin, CT, May 15, 1823; f. Henry; m. Marilla (Weldon); w. Elizabeth Harding; c. Edward H. Barnes. Early education in Philadelphia, PA, Tracy's School, New Britain, CT, at Burlington, NJ, and at an engr. school in Philadelphia. Commissioned 1st lt. of infantry in the 102nd Regiment of PA, 1844. Studied engr. in Europe in 1846. Appt. asst. engr. and made preliminary surveys of Western Division of the Pennsylvania RR, May 1, 1847. Principal asst. engr., made final location of line of the RR from Allegheny Mountains to Pittsburgh and const. the Western Division between same points, 1848-54. Chief engr. of Pittsburgh & Connellsville RR, 1854-57. Built branch line for Pennsylvania RR Co., Downington & Waynesburgh RR, 1858. Took charge of location of Dutchess & Columbia RR, 1860. Chief engr. in extension of Boston, Hartford & Erie, supt. const. on river terminal until financial difficulties caused the work to end in 1869. Chief engr. and promoter of CT Western RR Co., 1870. Pres. of co. in NYC authorized to const. underground RR from City Hall Park to Harlem. Prepared surveys and plans and began work, but the local political corruption made it impossible to finance the project. Appt. chief engr., presented plans of the NY Central Underground Co. to Rapid Transit commissioners. Other plans were used. Appt. consulting engr. of the proposed South Pennsylvania RR, 1882. Chief engr. of the NY, Lake Erie and Western RR and Coal Co., 1884. Built a line of RR from the Erie RR to the co.'s coal lands in PA. Mem. of New Croton Aqueduct Commission and chair of the const. committee, 1885-87. Became chief engr. of the NY and Long Island RR. Chief engr. of the NY Connecting RR; system to connect the NY, New Haven and Hartford RR with the Long Island RR system in Brooklyn and from South Brooklyn by ferry to the PA road in NJ. Mem: ASCE, Union League Club, New England Soc. (life mem.) Refs: WAB, WWW.

BASSETT, CARROL PHILLIPS; b. Brooklyn, NY, Feb. 27, 1863; d. Jan. 9, 1952; f. Allan Lee; m. Caroline (Phillips); w. Margaret Condit Kinney; c. Carrol Kinney, Estelle Condit, William B. K. Entered Lafayette College in 1879, grad. as valedictorian in 1883 with degree of civ. engr. Designed and const. the water works, sewerage and sewage purification works of many towns in NJ, NY, PA, CT, WV, and SC, 1886-1910. Pres. of NJ Sanitary Assn., 1892-93. Est. and operated water and electric companies, 1900-22. Const. engr., pres. Bassett Estates, Inc; Commonwealth Water and Light Co; Commonwealth Land Co; Lakewood Water & Coast Electric Co. and subsidiaries; vice pres. Summit Home Land Co; chair, Bd. of First National Bank and Trust Co.; dir. State Title and Mortgage Guaranty Co., Fireman's Insurance Co., and Commercial Casualty Co. Trustee of NJ Hist. Soc., Lafayette College, YMCA. Mem: ASCE, Am. Geographical Soc., New England Water Works Assm., NJ Soc. Colonial Wars, NJ Hist. Soc., NJ Washington Assn., Phi Beta Kappa, Tau Beta Pi, Phi Delta Theta (pres.). Publ: "The Conservation of Streams," "Inland Sewage Disposal" and other technical papers. Refs: WWW, WAB.

BATES, ONWARD; b. St. Charles County, MO, Feb. 24, 1850; d. Augusta, GA, April 4, 1936; f. Barton; m. Caroline Matilda (Hatcher); m. Virginia Castleman Breckinridge; c. none. Attended local schools in St. Charles County. At age 15 went to St. Louis, became apprentice in the Fulton Iron Works. Engaged first as draftsman then as inspector by Charles Shaler Smith on bridge over Missouri River at St. Charles. Spent two years at RPI. Returned to St. Louis in 1873 and worked on the Eads bridge serving as associate of one contractor and as inspector on steel and iron work for St. Louis Bridge Co. until 1874. Draftsman with Cincinnati Southern RR, became inspector of iron bridges and trestles. Similar position with Chicago, Milwaukee & St. Paul RR, 1877. Went to Australia in 1878, designed and built several iron bridges with the Edge Moor Iron Co. of Wilmington, DE. Returned to US in 1883, served as pres. of the Pittsburgh Bridge Co. Mining venture in Mexico, 1887-88. Returned to Chicago, Milwaukee & St. Paul as engr. and supt. of bridges and buildings. Organized Bates and Rogers Const. Co., 1901-1907. Hon. degrees from Univ. of Wisconsin (1897), RPI (1918) and Univ. of Missouri (1924). Mem: ASCE (vice pres., 1906-1907, pres., 1909, hon. mem., 1923), Western Soc. of Engrs. (pres.),

Engrs. Club of NY, City Club of Chicago, Inst. of Civ. Engrs. of Great Britain, Royal Soc. of the Arts. Refs: DAB, WAB, WWW, WWE.

BAYLES, JAMES COPPER; b. NYC July 3, 1845; d. May 7, 1913; f. James; m. Julia Halsey (Day); w. Ianthe Green; c. Lewis C. and Howard J. Educated in public schools, enlisted in the Union Army. Entered 22nd Regiment of NY, 1862, commissioned an artillery lt. After one year illness, worked for the Delamater Iron Works of NY. On engr. staff of the Raritan & Delaware Bay RR. Acting editor of the *New York Citizen* during illness of the editor, 1865. Editor of the *Commercial Bulletin*, 1868-1869. Resigned to accept the editorship of *Iron Age*, 1869-89. Founded trade paper, *The Metal Worker*, acted as first editor, 1876. Lectured on sanitation and water supply. Pres. of NJ State Sanitary Assn., 1884. Commissioner to plan and const. new sewer system in NJ. Non-resident lecturer on labor problems at Sibley Engineering College of Cornell, 1886. Pres., Bd. of Health of NYC, 1887-88. Mem. of the editorial staff of the *New York Times* and engaged in consulting engr. practice in connection with city depts. and public utilities. Mem: Am. Inst. of Mining Engrs. (pres., 1884-85), charter mem. of Am. Soc. of Mech. Engrs. Publ: *House Drainage and Water Service in Cities, Villages, etc., With Considerations of Causes Affecting the Healthfulness of Dwellings* (1878), *The Study of Iron and Steel* (1884), *Causes of Industrial Depression* (1884), *Industrial Competition* (1885), *Iron Manufacture in the Southern States* (1885), *The Engrineer and the Wage Earner* (1885), *Professional Ethics* (1886), *The Shop Council* (1886), and many papers to transactions of engr. societies. Refs: DAB, WAB, Lamb.

BEAHAN, WILLARD; b. Watkins, NY, Jan. 15, 1854; d. Feb. 5, 1928; f. James; m. Harriet (Griswold); w. Bessie Bell Dewitt, 1892; c. James Dewitt. Grad. from Starkey Seminar, Eddytown, NY, 1873. Degree of civ. engr. from Cornell Univ., 1878. Entered US Army Corps of Engrs. as asst. instrumentman and as computer on the Mississippi River Survey, 1878. Rodman on Texas & Pacific RR of the Gould Lines in the southwest, 1880. Position of division engr. in 1881, then became division engr. on Fort Worth & Denver City RR in charge of location and const. Transitman on the Lehigh Valley RR, March 1883-March 1884. Division engr. on the Missouri Pacific RR, engaged in location, const. and maintenance until Feb. 1889. Chief of const., Chilean RR, South America, 1891. Supt. of Streets, St. Louis, MO, 1891. Worked for Anderson and Barr, Contractors, Nov. 1891-Dec. 1896. Lecturer on RR location at Cornell and Leland Stanford, Jr. Universities. Supt. const. for James J. Hill at Cascade Tunnel, Aug. 1897-May 1898. Principal engr. for Lehigh Valley RR, 1899-1901. Division engr., Chicago and North Western RR, 1901-1904. First asst. engr., NY Central RR at Cleveland Ohio, 1905-24. Special engr. in charge of relocation surveys on the NY, Chicago and St. Louis RR, 1924-26. Trustee, Cornell Univ. Mem: ASCE (dir., 1919-21), Bureau of Municipal Research, Cleveland, OH, Cleveland Engr. Soc., Bd. of Civil League of Cleveland, Cleveland Chamber of Commerce, Dir. of the Bureau of Municipal Research, Am. Railway Engr. Assn. Publ: *The Field Practice of Railway Location*. Refs: WWW, TASCE, WWE.

BEARDSLEY, JAMES WALLACE; b. Coventry, NY, Sept. 11, 1860; d. May 15, 1944; f. William Hurd; m. Catherine Tremper (Phillips); w. Ellen J. Pearne; c. Wallace Pearne. Grad. State Normal School, NY, 1884. Degree of civ. engr. from Cornell Univ., 1891. Asst. engr., city of Newton, MA, 1891-92.

Asst. engr. with Sanitary District of Chicago in charge of const., 1892-98. With US Bd. of Engrs. on Deep Waterways in charge St. Lawrence River Surveys, 1898-1900. With US Army Corps of Engrs. in charge of Fox River surveys and Sand Beach Harbor improvement works, 1900-1902. Consulting engr. to Philippine Islands Commission, 1902-1903. Chief, Philippine Bureau of Engr., 1903-1905. Dir. of Public Works, Philippine Islands, 1905-1908. Consulting engr. investigating irrigation in Java, India and Egypt, 1908-1909. Irrigation engr. with J. G. White & Co., NYC, 1909-10. Chief engr., Puerto Rico. Irrigation Service, 1910-16. Private consulting practice, 1916-18. Consulting engr. in ordinance, 1918. Asst. chief engr., Grand Canal surveys of China, 1918-19. Chief engr. and mem., Junta Central de Caminos, Panama, 1920-21. Consulting engr. in private practice, 1922-25. With Obras Publicas, Santo Domingo, Dominican Republic, 1926-29. Private consulting practice, Syracuse, NY, 1930- . Mem: ASCE, Western Soc. of Engrs., Delta Phi, Univ. and Columbos Clubs of Manila, Cornell Club of NY)C, Cosmos Club of Washington, DC, Sphinx Head, Cornell Univ. Refs: WWW, WAB, WWE.

BENHAM, HENRY WASHINGTON; b. Quebec, Canada, April 8, 1813; d. NYC, June 1, 1884; f. Jared Benham; m. Rebecca Hill; w. Elizabeth McNeill; c. 2 daughters and 1 son. Moved to CT when mother remarried. Entered Yale College., 1832. Appt. to US Military Academy, July 1, 1833 and grad. 1st in class, 1837. Brevetted 2nd lt. of engrs., July 1, 1837. Engaged in engr. duties in connection with coast defenses. Asst. in charge of improvements in Savannah River. Promoted 1st lt., July 7, 1838. Supt. engr. of repairs of Fort Marion and sea-wall at St. Augustine, FL, 1839-1844. Repairs of the defenses of Annapolis Harbor, 1844-45, and in 1845 resumed work on the sea wall, remained until 1846. Employed during this time on repairs at Fort Mifflin, PA and at Fort McHenry, MD and other govt. works. Served in Mexican War, wounded at battle of Buena Vista, Feb. 22-23, 1847. Promoted to capt., May 24, 1848. Repairs for defenses of NY Harbor, and sea wall const. in Boston Harbor. Built Buffalo Lighthouse, 1852, and Washington Navy Yard, 1852-53. Built Boston Lighthouse. Asst. in US Coast Survey Office, March 29, 1853-Nov. 1, 1856, sent to Europe in connection with these duties. Supt. engr. of const. of Forts Winthrop and Independence, Boston Harbor, 1856-57. Supt. survey of site for fort at Clark's Point, New Bedford, 1858. Charge of repairs of Fort Adams, Newport, RI, 1857-58. Supt. building of fort at Sandy Hook, NJ, 1858-61, and Potomac aqueduct, 1860. Engr. of Quarantine Commission of Port of NY from 1859-60. At beginning of Civil War, appt. chief engr. Dept. of Ohio, US Army, May 14, 1861. Commanded maj. of engrs., US Army, Aug. 6, 1861. Brevetted col. for gallantry at the battle of Carrick's Ford, VA, July 13, 1861. Commanded. brig. gen., US Volunteers, Aug. 1861. Engaged in Virginia Campaigns, including actions at New Creek, Aug. 16, 1861 and Carnifax Ferry, Sept. 10, 1861. Charged with violation of orders during Battle of Secessionville, SC, 1862. Appt. of brig. gen. revoked, Aug. 7, 1862. Supt. fortifications at Boston and Portsmouth Harbors, in command of northern district of Dept. of the South. Revocation of promotion cancelled by Pres. Lincoln, Feb. 6, 1863. Promoted to rank of lt. col. of engrs. in regular Army, March 3, 1863. Built pontoon bridges across the Rappahannock, Potomac and James Rivers. Commanded engr. brigade, Army of the Potomac, 1863. In charge of Pontoon Dept., Army of the Potomac, 1864. Brevetted maj. gen. US Volunteers, 1865. Commanded col. engrs., 1867. In charge of defenses of Boston, NY Harbors. Promoted col. of engrs., March 7, 1868. Invented picket shovel and rapid const. of pontoon bridges by means of "simultaneous bays." Retired from service, 1882. Refs: CAB, Lamb, DAB, WWW, WAB, DR, BDNA.

JOHN ANDERSON BENSEL (1863-1922), PRESIDENT OF ASCE, 1910

BENSEL, JOHN ANDERSON; b. NYC, Aug. 16, 1863; d. Bernardsville, NJ, June 19, 1922; f. Brownlee; m. Mary Maclay (Hogg); w. Ella Louise Day; c. Louise Day, John A. Jr., Evelyn Adelaide and Henry Day. Educated in public and private schools in NY. Grad. Stevens Inst. of Technology, degree of mech. engr., June 12, 1884, hon. D. Engr., June 21, 1921. Rodman in Dept. of Public Works in surveys for the NY Aqueduct and new Croton Dam, 1884. Rodman in the Maintenance of Way Dept., Oct. 1, 1884, asst. engr., 1886, and asst. supervisor, Pennsylvania RR, Sept., 1888. Asst. engr. and asst. supervisor in charge improvement of dock and freight terminals, Pennsylvania RR, 1887-89. Asst. engr. Dept. of Docks, NY in charge const. work on bulkhead, sea-wall docks and the supervision of private work, North River water front, 1889-96. Took up private practice, consulting engr. for Central RR of NJ in the valuation of their dock property. Consulting engr. for the Girard estate, Philadelphia, in const. for the river wall along a mile of water-front on Delaware River, 1895-98. Designed other piers and water front structures for private parties on Delaware River. Member of firm of Steers and Bensel, 1896-98. Dept. of Docks and Ferries, NYC, asst. engr. 1896-97, engr. in chief, 1898-1906. Cmdr. of docks and ferries, 1907. pres., Bd. of Water Supply, NYC, 1908-10, elected 1910, re-elected 1912, serving 1911-14, State engr. of NY, conducting many works of public improvement including the Barge Canal. Maj. gen. US Army Corps of Engrs. commanding 125th Engr. Batallion, 1917-18. Consulting engr. NY and NJ Bridge and Terminal Commission and NJ Interstate Bridge and Terminal

Commission. Mem: ASCE (dir., 1899-1901, vice pres., 1907-1908, pres, 1910), Am. Soc. of Mech. Engrs., Am. Inst. of Mining and Metallurgical Engrs., Inst. of Civ. Engrs. of Great Britain, Lawyers Union and Univ. Clubs of NYC, Fort Orange Club of Albany, NY, golf clubs and country clubs. Publ: "Observations on Dock Work in NY Harbor," "Final Report of the Special Committee to Investigate the Conditions of Employment of and Compensation of Civil Engineers," "Final Report of the Special Committee on Floods and Flood Prevention" and "Address at the 42nd Annual [ASCE] Convention, Chicago, Ill., June 21, 1910". Refs: WWW, WAB, WWE, PASCE.

BENZENBERG, GEORGE HENRY; b. NYC, May 31, 1847; d. May 31, 1925; f. Henry ; m. Christina (Rugee); w. Alvina Wolfrum; c. May. Parents moved to MI when one year old. Primary education in public schools and German-English Academy. Entered Univ. of Michigan, 1863, grad. with degree of civ. engr., June 1867. Asst. engr., US Lake Survey, 1867-69. Transitman, IA Division of Chicago, Milwaukee & St. Paul RR, 1869. Designed and const. iron ore shipping docks at Milwaukee for Bayview iron works and also track system of the plant. In charge of location and const. of part of Milwaukee and Northern RR, 1871. Asst. city engr. of Milwaukee 1874-82. City engr. in 1882-99. Entire management of Water Dept. and pres. of Bd. of Public Works. Some accomplishments in Milwaukee: improved street pavements; built iron bridges, viaducts, some of the first bascule bridges, relief sewers; water meters introduced, designed and const. a new water work tunnel, const. new high service pumping station. Consulting engr. to Chicago, Kansas City, Cleveland, Toledo, New Orleans, Cincinnati and private corporations. Hon. D.Engr., Univ. of Michigan, 1912, Sc.D., Univ. of Wisconsin, 1911. Mem: ASCE (dir., 1895-1897, vice pres., 1901-1902, pres., 1907-1908) Am. Water Works Assn. (pres., 1893-94), Am. Soc. of Municipal Improvements (pres., 1894-96), New England Water Works Assn., Soc. for the Promotion of Engr. Education, National Geographic Soc., Am. Forestry Assn., Am. Assn. for the Advancement of Science, Mason, Bd. of Trustees of the Northwestern Mutual Life Insurance Co. Refs: WWW, WAB, WWE, TASCE.

BERG, WALTER GILMAN; b. NYC, Jan. 12, 1858; d. NYC, May 12, 1908; f. Albert W.; m. Helen McGregor (Morse); w. Ruby Burke; c. 2 sons. Educated in public schools of NYC. Taken to Europe when he was nine years old, educated in Germany. Grad. from Royal Polytechnic Inst. at Stuttgart, Germany, degree of civ. engr., 1879. Awarded the Gold Medal by the King of Wurtemberg for treatise on "Spherical Conic Sections," and a scholarship at the Royal Polytechinic Inst. Returned to US in 1879, draftsman and shop inspector for the Delaware Bridge Co. until 1880. Appt. engr. of bridges with the Richmond and Allegheny RR Co. Entered the service of East Tennessee, Virginia, Georgia RR Co. as principal asst. engr. of the const. dept., 1882. Asst. engr. with Lehigh Valley RR Co. in charge of Lehigh and New Jersey Division on design and const. of shop buildings, round-houses, surveys for new branches. Charge of office of the chief engr. Designed and const. the Lehigh Valley Creosoting plant at Perth Amboy, NJ. In charge of the operation of this plant, among the first of its kind in the US. Appt. principal asst. engr. of the Lehigh Valley RR with headquarters at Jersey City, NJ and designed and const. the first piers at the latter place. Engr. of maintenance of way of Lehigh Valley RR with headquarters at South Bethlehem, PA. In charge of the Engr. Dept. until Feb. 1,

1900. Appt. chief engr. of the RR in charge of const. until his death. Designed and const. the RR Shop System at Sayre, PA. Appt. as representative of the Am. Railway Engr. and Maintenance of Way Assn., died before he left for Washington, DC. Mem: ASCE, Am. Railway Engr. Assn. (6th pres.), Assn. of Railway Supt. of Bridges and Buildings (pres.), Am. Roadmasters Assn., Eastern Maintenance of Way Assn., Railway Supts. Assn., Am. Soc. for Testing Materials, New York RR Club, Founders and Patriots of America and Holland Lodge, F. and A.M. Publ: *Buildings and Structures of American Railroads* (1892), *American Railway Bridges and Buildings* (1898), *Timber Tests* (1899), *Railway Shop Systems* (1904). Refs: TASCE, WWW.

BIGELOW, CHARLES H.; b. Watertown, MA, July 13, 1814; d. New Bedford, MA, April 15, 1862; f. Tyler; m. Claissa (Bigelow); w. Harriet Briggs. Grad. US Military Academy, 1835. Served in the Corps of Engrs., 1835-46. Asst. engr. for const. of Ft. Warren and Ft. Independence in Boston Harbor. Resigned commission on June 16, 1846 with the rank of capt. Chief engr. of Essex Co. in const. of the city of Lawrence, MA, 1846-57. Supervised const. of the dam, North Canal, Lawrence Machine Shop, and the Atlantic, Pacific, Duck and Pemberton Mills. Consulting engr. to Lewiston, ME Water Power Co., 1850-51, the Niagara Falls Canal, 1852, and helped prepare plans for a projected dam and canal at Sherbrooke, Quebec, Canada, 1856. Engr. of water measurement for the Augusta, Georgia Manufacturing Co. and Minneapolis Mills Co. dam and canal at the Falls of St. Anthony. Civ. engr. in fortification of New Bedford Harbor, 1859. Served as supt. and engr. of the New Bedford Copper Co., 1860-61. In charge of fort at Clarke's Point, New Bedford, 1861 and 1862. Refs: Poirier.

BLACK, WILLIAM MURRAY; b. Lancaster, PA, Dec. 8, 1855; d. Washington, DC, Sept. 24, 1933; f. James; m. Eliza (Murray); w. Daisey Peyton; w. 2nd Gertrude Totten Gamble; c. Roger Derby, Percy Gamble, William Murray. Educated at high school in Lancaster, and Franklin and Marshall College, 1870-73. Appt. to US Military Academy by competitive exam; grad. June 14, 1st in class, 1877. Commissioned 2nd lt., Corps of Engrs. Served at US Military Academy as asst. instr. of practical military engr. Entered the Engr. Battalion at Willets Point, Dec. 21, 1877; grad. in Engr. School of Application March 1, 1880. Instr. in practical military engr. at US Military Academy, 1882-86. Instr. in civ. engr. at Willets Point, 1891-95. Practical const. work on locks and dams on Kanawha and Ohio Rivers. Secy. of Harbor Commission in Philadelphia, PA. District engr. in FL District, opened mouth of the St. John's River. While in FL, initiated use of reinforced concrete in fortification const., contributed exact knowledge on littoral drift of sand, behavior of bars in tidal estuaries, and handling concrete in sea water. Asst. in charge of fortifications, Office of Chief of Engrs., Washington, DC, 1895-97. Engr. commissioner of District of Columbia, March 1897-98, prepared plans for abolition of RR grade crossings in the District and secured their adoption. Promoted 1st lt., 1880; capt., 1887; maj. and lt. col. of Volunteers in Spanish-American War, 1898. Served as chief engr. in Puerto Rico campaign and commanded landing party at Guanica, July 25, 1898. Chief engr. of Dept. of Havana, Jan. 2, 1899-April 30, 1900. Chief engr. in Division of Cuba, Jan. 1900-April 1901. Est. public works department, starting projects for sewers, paving and an improved ocean front. Supervised engr. work in other parts of Cuba and made military survey of the island. Commanded 3rd Battalion, Engr. School of the Army, and post of Washington Barracks, D.C. from 1901-1903. Observed work of Isthmian Canal Commission in Panama, April 1903-July 1904. In charge of rivers, harbors and fortifications in ME, 1904-1906. Advisor, Dept. of Public Works in Cuba during second occupation, 1906-1909. Stationed in NY, Feb. 1909-March 1916, in charge of improvement of East River, Hell Gate and Hudson River and its tributaries. Received Sc.D., Franklin and Marshall College, 1912. Senior member of bd. to remove wreck of U.S.S. *Maine* from Havana Harbor and determine cause of its destruction, 1910-1913. Plan for intracoastal waterway and const. of memorials of victory of Commodore Macdonough on Lake Champlain. Senior member of Bd. of Engrs. for Rivers and Harbors, and of NY and Boston Harbor Line Bds. Promoted to lt. col., 1905; col. in 1908; brig. gen. and Chief of Engrs. in March 1916. Duties included connection with Mexican Border and incorporation of practical RR men into the Army. Mobilized and shipped ten regiments of RR engrs. to France. Enlarged Corps of Engrs. more than 100 times, organized engr. reserve corps and est. Ft. Humphreys, VA, as replacement center and site for army engr. school. Supervised work of dir. gen. of US Military RRs. Served as mem. of National Research Council and as mem. of Committee on Engr. and Education, Council of National Defense. Chair, Inland Waterway and Railway Transportation, Commission of Council of National Defense, 1917. Accompanied Secy. Baker on inspection of the Army in France, 1918. Retired Oct. 31, 1919, with rank of maj. gen. Awarded the Distinguished Service Medal for planning and administering engr. and military RR services. Received Ph.D. in engr., PA Military College, 1920. Consulting engr., Emergency Fleet Corp, 1919-1920, Intl. Whangpoo Conservatory Commission, Shanghai, 1921, and his own firm, Black, McKenny & Stewart, 1920-29. Firm designed and supervised const. of extensive improvement at the mouth of Magdalena River in Republic of Columbia. Invented method of purifying sewage by aeration. Mem; ASCE, National Research Council, Am. Legion, Am. Assn. for the Advancement of Science (fellow), Soc. of Am. Military Engrs. (1st pres.), Washington National Monument Soc. Publ: "The Improvement of Harbors on the South Atlantic Coast of the United States" (Thomas Fitch Rowland Prize, ASCE, 1893), "Public Works of the United States" (1893), and "Report on Discharge of Sewage into New York Harbor" (1910), "Waterways and Railway Equivalents" (Arthur M. Wellington Prize, ASCE, 1923). Refs: DAB, WWW, TASCE, WAB, WWE.

BOGART, JOHN; b. Albany, NY, Feb. 8, 1836; d. April 25, 1920; f. John Henry; m. Eliza (Hermans); w. Emma Cherrington Jefferis. Educated at the Albany Academy, awarded the Van Rensselaer Classical Medal and the Caldwell Mathematical Medal. Attended Rutgers College, grad. with A.B. degree in 1853, M.A. in 1856, Sc.D. 1912. Engr. Corps, NY Central RR, state canals of NY. Asst. in the engr. dept. of the state of NY, employed on the reconst. and enlargement of the canals in the Eastern Division. Engaged in the const. of Central Park in NY on roads, bridges, tunnels, drainage and water system, until the beginning of the Civil War. Engr. during the Civil War at Ft. Munroe, VA, and other points, 1861-66. Chief engr. Park Commission, Brooklyn. Chief engr. Dept. of Public Works, NYC, 1872-77. Worked on public and municipal works at Albany, New Orleans, Chicago, Nashville, Baltimore, Buffalo, Norfolk, Kansas City, Toronto, Keene, Rochester, 1877-86. Deputy State Engr. and Surveyor of NY State, 1886-Summer 1887. Elected State Engr.

and Surveyor, Jan. 1, 1888. Consulting engr. of Washington Bridge, NY, 1887. Consulting engr. for the Cataract Const. Co., traveled all over Europe studying existing methods of power generation and transmission. At Domene in the Dauphine Alps, found precedent for system adopted at Niagara Falls. Chief engr. of the Chatanooga and Tennessee River Power Co. Engr. hydraulic and electric dev. of power at Sault Ste. Marie, Massena, St. Lawrence River, Cascade, British Columbia, Knoxville, Atlanta, AZ, southern VA, NE and Youghioghney Power Co. Consulting engr. for several RRs. Advisory engr. for the original Rapid Transit Commission of NY, prepared plans and contracts for the first subway system. Prepared plans for tunnels under the Hudson to Jersey City and Hobooken and for the subway now operating between NY and Queens. Served on: Nicaraguan Canal Commission, RR Terminal Commissions of Buffalo, NY and Toronto, Ontario, Canada, Croton Aqueduct Commission, NY State Bd. of Health. Represented US at the Intl. Navigation Congresses held in Dusseldorf, Germany in 1902, Milan, Italy in 1905 and St. Petersburg, Russia in 1908. Only civilian mem. of bd. of engrs. to examine feasibility of const. deep waterway from Chicago, IL, to the Gulf of Mexico. Served as Chair of the Inland Section of the Congress, 1912. Formed firm of Bogart and Pohl, 1913. Mem: ASCE (dir., 1873-1875, treas., 1876-77, 1891-94, secy., 1878-1890), Municipal Art Commission of NY, Am. Inst. Architects (hon. mem.) Publ: *Engineering Feats; Papers and Discussions*. Refs: DAB, WWW, TASCE.

BOND, EDWARD AUSTIN; b. Dexter, MI, April 22, 1849; d. Dec. 10, 1929; f. Hollis; m. Emily (Faxon); w. Gertrude Hollenbach; w. 2nd, Clara E Ellis; w. 3rd Mrs. Elizabeth Parsons. Educated in public schools in MI and business college of Utica, NY. First engr. work with the Delaware, Lackawanna & Western RR, 1867-70. Resident engr. of the Utica & Black River RR, 1870-71, chief engr., Clayton and Theresa RR 1871-75, Utica & Black River, 1875-89. Chief engr. and gen. supt. of the Carthage & Adirondack RR from Carthage to Benson Mines and the Oswegatchie River, 1886-89. Mem. of Hinds & Bond of Watertown, NY, gen. engr. and contracting in RR work and the const. and operation of water works in the US and Canada, 1889-96. Engaged in private practice, July 1896- Nov. 1898, as consulting engr. on various works. Elected State Engr. and Surveyor of NY, Nov. 1898-May 1, 1904. Commissioner of the Land Office, a member of the Canal Bd., Bd. of State Canvassers, State Bd. of Equalization of Assessments, and prior to their reorganization, was a member of the State Bd. of Health and Forest Preserve Bd. Chair, advisory bd. of consulting engr. for const. of Barge Canal, 1904-July 21, 1911. Private practice on canal and harbor works, visited Uruguay on a harbor project in 1914. Mem., advisory bd., Trans-Alaska and Siberia RR. Pres. Barrie, Ontario, Canada Water Works Co., Chatham, Ontario, Canada Water Works Co. Napenee (Ontario, Canada), Water Works Co., Mason Mem: ASCE, Mason, Lincoln League. Refs: TASCE, WWW, WAB, WWE.

BORDEN, SIMEON; b. Freetown (Fall River), MA, Jan. 29, 1798; d. Fall River, MA, Oct. 28, 1856; f. Simeon; m. Amy (Briggs); unmarried. Moved to Tiverton, RI, 1806, educated in the public schools there. Began work in Pocasset Machine Shop 1826, supt., 1828. Const. base bar used for measuring base line in trigonometrical map survey of Boston, 1830. Assisted in survey, 1831-41, the first geodetic survey in the US. Chief surveyor in 1834. Mem. MA legislature, 1832-33, 44-45, 49. Made survey of line between RI and MA used in boundary dispute case, 1844. Chief surveyor RRs in ME, NH, MA, CT, 1841-51. Often called upon as expert in the courts. Connected palisades on one side of Hudson River with Ft. Washington on the other by overhead telegraph lines, 1851. Mem: Academy of Arts and Sciences, Philosophical Soc. Publ: *Engineer's Report of the Worcester and Keene Surveys* (1846), *A System of Useful Formulae, Adapted to the Operations of Locating and Constructing Railroads* (1851; based upon paper read before the Boston Soc. of Civ. Engrs., Dec. 1849). Refs: Lamb, WWW, DAB, CAB, WAB, DR, BDNA.

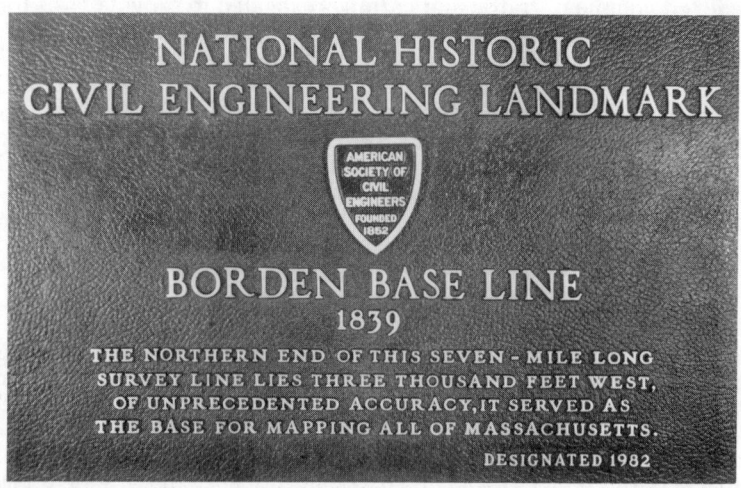

LANDMARK PLAQUE FOR THE BORDEN BASE LINE, NAMED FOR SIMEON BORDEN (1798-1856)

BRINCKERHOFF, HENRY MORTON; b. Beacon, NY, April 20, 1868; d. 1949; f. Peter Remson; m. Helen Morton; w. Florence Louise Fay; c. Henry Morton and Remson. Grad. from Trinity School in NYC. Privately tutored in advanced mathematics. Entered Stevens Inst. of Technology, grad. valedictorian of class with degree in mech. engr., 1890. Worked for one year on electrification of horse car lines of Boston St. RR Co. and one year on installations for General Electric Co. Asst. engr. of Utica NY Belt Line Street RR Co., 1892. Asst. engr. for intramural RR at World's Columbian Exposition, Chicago. Invented, installed and patented third rail system, 1893. Remained in Chicago after the Exposition, sucessively an electrical engr., 1894, supt. of motive power and way, 1895, and asst. gen. mgr. of Metropolitan West Side Elevated RR Co., 1897. Gen. mgr. of Metropolitan West Side Elevated RR Co., 1899, first permanent RR to install third rail system. Visited RR systems in Europe. Resigned in 1906 to become partner of William Barclay Parsons and Eugene Klapp in firm of Barclay, Parsons & Klapp, NY, consulting engrs. Senior member of successor, Parsons, Brinckerhoff, Hogan and MacDonald. Directed survey of Detroit street

RRs, 1913-1919. Similar work for Cleveland Rapid Transit Commission, 1919-20. Suggested plans for relieving congestion and for installation of subway system. Appt. chief engr. of Chicago traction and subway commission to investigate conditions and to determine plan for consolidated operation of Chicago's elevated, surface and proposed subway lines, Feb. 1916. Engr. during World War I with Emergency Fleet Corp. on const. of dry docks at Norfolk Navy Yard and with Federal Housing Corp. Supervised building of ferries and misc. structures at Portsmouth and Norfolk, VA. Following war, made valuations of more than $1 million including Philadelphia Gas Co., Chicago surface and elevated RR lines, Detroit street RR and suburban lines and Detroit City Gas Co. Predicted volume of traffic and earnings expected to provide basis for setting bond issue to support const. of Pennsylvania Turnpike, 1927. Also contributed design for the road. Appt. member of Chicago subway commission representing federal public works administration on design and const. of Chicago subway system, 1935. Chair, Englewood Planning Bd., 1935-40. Appt. to seat on Chicago Subway Commission by Public Works Administration, 1938. Vice pres., Parkland Const. Co., built Detroit Vehicular Tunnel. Vice pres.,Parklap, Inc., contracting co. Prepared plans and specifications for transportation for NY World's Fair 1939, Inc. Mem., Cincinnati Transit and Internal Bus and Tractor Commission, 1940. Conducted investigation of prospects of rehabilitating NY, Boston & Westchester RR for presentation to court in receivership litigation, 1941. During World War II, served Army and Navy as dry-dock engr. and member of Carribean Architect Engrs. on govt. defense work. Mem., NY State Thruway Commission reporting on traffic estimates, 1945. Formally retired in 1946, served as consulting assoc. until his death in 1949. Mem: ASCE, Am. Inst. of Electrical Engrs., Inst. of Traffic Engrs., Western Soc. of Engrs., Theta Xi and the Down Town Assn. of NYC. Refs: WAB, YPB, WWE.

BROADHEAD, GARLAND CARR; b. near Charlottesville, VA, Oct. 30, 1827; d. Dec. 12, 1912; f. Achilles; m. Mary Winston (Carr); w. Marion Wallace Wright; w. 2nd Victoria Regina Royall; c. 2 sons and 2 daughters. Education in private schools and under tutors in St. Charles County, MO. Attended Univ. of Missouri, 1850-51, hon. M.S., 1873. Western Military Inst. of Kentucky, 1851-52. Civ. engr. Missouri Pacific RR, 1852, asst. engr. in charge of location lines, resident engr. of const., 1857. Asst. geologist of MO, 1857-61, made geological reconnaissance along line of the southwestern branch of Pacific RR. Asst. adjutant gen. during Civil War, US Deputy collector internal revenue, Missouri, 1862-64. Asst. engr., Missouri Pacific RR, 1864-66. US Assessor 5th District MO, 1866. Asst. geologist of IL, 1868, of MO, 1871-73, state geologist of MO, 1873-75. Surveys and const. of RRs in Kansas, 1879-80. Special agent 10th census on quarry industry for MO and KS, 1881. Nov. 1883-April 1884 engaged in arranging specimens in the museum of the state univ. at Columbia, MO. Served on the Missouri River Commission, July 1884. Visited Yellow Stone Park with members of the commission and the upper streams tributary to the Missouri, July 1885. Bd. of Mgrs., Bureau of Geology and Mines of Missouri. Mem: Bd. of Jurors, Centennial Exhibition, Philadelphia, 1876. Prof. geology, Univ. of Missouri, 1887-97. Mem., Intl. Geological Congress of Geologists. Mem: Geological Soc. of America, Am. Assn. for the Advancement of Science, National Geographic Soc., Am. Inst. of Mining Engrs., Am. Forestry Assn., St. Louis Academy of Science, Kansas City Academy of Science, Davenport Academy of Science, Virginia, Illinois and the Missouri Hist. Socs. Author of geological reports on MO and IL, and other geological publications. Refs: WWW, CAB, WAB, DAB, BDNA.

BROWN, CHARLES CARROLL; b. Austinburg, OH, Oct. 4, 1856; d. Nov. 26, 1949; f. George Pliny; m. Mary Louise (Seymour); w. Cora Stanton; w. 2nd Eileen Finkle; c. Edwin Stanton. Studied engr. at Cornell Univ., 1874-75. Grad., Univ. of Michigan, 1879, degree of civ. engr. Hon. degree, A.M., 1913. Asst. engr., US Lake Survey, 1879-83. Asst. NY State Survey, 1884. Consulting engr., mainly in municipal sanitary and highway lines, 1885- . Prof., civ. engr., Rose Polytechnic Inst., 1885-86. Prof., civ. engr. in charge of dept., Union College, 1886-93. Consulting engr. NY State Bd. of Health, 1887-93. City engr., Indianapolis, IN, 1894-95. Asst. city engr., Omaha, NE, 1896. Editor, *Municipal Engineering Magazine*, 1896-1917, becoming treas., vice pres. and gen. mgr. of the co. With IL Division of Highways, 1918-19. Prof., civ. engr., Valparaiso Univ., 1919. Dean of College of Applied Sciences, 1920-21. Treas., Public School Publishing Co., engr. Dept. of Public Works, St. Petersburg, FL, 1921-23. City engr., Lakeland, FL, 1923-27. Prof., civ. engr., Univ. of Florida, 1927-33. Mem: ASCE, AAE, Sigma Xi, Am. Water Works Assn., Am. Soc. for Municipal Improvements, Indiana Engr. Soc., Illinois Soc. of Engr. and Surveyors, founder of National Assn. of Cement Users. Publ: *Report on Croton Water Shed of NYC* (1889), *Directory of American Cement Industries* (5 editions, 1901-1909), *Handbook for Cement Users* (3 editions, 1901-1905). Refs: WWW, WWE.

BROWN, WILLIAM HENRY; b. Little Britain Township, PA, Feb. 29, 1836; d. visiting Belfast, Ireland, June 25, 1910; f. Levi K.; m. Hannah C. (Moore); w. Sarah A. Rimmell. Educated in public schools and Central High School in Philadelphia. Taught himself surveying. Teacher at Port Deposit, MD, 1856-58. Asst. engr., Survey Dept., Philadelphia, 1858; helped lay out first street RR. Asst. engr., US Military RRs, 1861-63. In charge of Orange, Alexandria & Manassas Gap RR, est. records in rapid const. of bridges and reconst. of the road. Asst. engr., 1863-64, principal asst. engr. Jan.-Oct. 1864, Pan Handle RR. Asst. engr., Pittsburgh Division, 1864-65. Engr., Oil Creek RR, March 17-July 1, 1865. Principal asst. engr., Philadelphia & Erie Division, 1864-65, then engr., 1867-69. Engr. in charge const., Altoona shops, 1869-70. Resident engr., Middle Division, 1870-71. Chief engr. and supt., Lewiston Division, 1871-72. Supt. Bedford Division, 1872-74. Engr. Maintenance of Way, 1874-81. Chief engr., July 1, 1881-Feb. 28, 1906, Pennsylvania RR. With the Pennsylvania RR, built 14 elevated RRs, 412 tunnels, 163 stone bridges including Rockville Bridge, one of the longest in the US at the time and largest in the world of stone. Mem: Pennsylvania Soc. of the Sons of the Revolution. Refs: WWW, DAB, WAB.

BUSH, LINCOLN; b. Palos Township, IL, Dec. 14, 1860; d. East Orange, NJ, Dec. 10, 1940; f. Lewis; m. Mary (Ritchey); w. Alma R. Green; c. Cedric Lincoln and Denzil Sidney. Grad. with M.A. at the Cook County Normal School in 1881. Taught for three years in rural schools. Univ. of Illinois, B.S., 1888, hon. D.E. 1904. Asst. engr., WY Division of the Union Pacific RR on maintenance and const. of the Pacific Short Line RR, 1888-1890. Jan. 1890, asst. instr. of mathematics at the Univ. of Illinois. Asst. civ. engr. with Morrison & Corthell bridge builders, April 1890-91. Chief Draftsman and office asst. engr. for the Pittsburgh Bridge Co., Dec. 1891-June 1896. Asst. engr. on masonry design for bridges with the Chicago Drainage Canal Bd., June 1896. Asst. bridge engr. and acting division engr. for the Chicago & Northwestern RR, Jan. 1897-99. Delaware Lackawanna & Western RR, principal asst. engr., Jan. 1900-1903, chief engr., 1903-1909. Opened office in NYC as a consulting engr., Jan. 1909. Vice pres. and chief engr. of the F. M. Talbot Co., Contractors, 1910-17. Pres. and chief engr. of Talbot Const. Co., 1911-13. Chief engr. of contracting firm of Flickwir & Bush, Inc, 1912-16. Pres. of Bush, Roberts and Schaefer Co., 1920-25. Served in Quartermaster's Corps, US Army in World War I with rank of col. Several inventions including the Bush Train Shed (1904) and Bush Track Construction (1908). Mem: ASCE (hon. life mem., dir., 1912, treas., 1915-16, pres., 1928), Am. Railway Engr. Assn., Western Soc. Engrs., Engrs. Club of NYC. Refs: WWE, DAB, WAB.

CAIN, WILLIAM; b. Hillsboro, NC, May 14, 1847; d. Dec. 7, 1930; f. William; m. Sarah Jane (Bailey); unmarried. Private school in Hillsboro, entered Hillsboro Military Academy, 1859. Drilled Confederate troops when only 14 yrs old. Studied law under Hon. J. L. Bailey at Asheville, 1865-66. Received M.A., 1866 from North Carolina Military Polytechnic Inst. Asst. engr. to W. C. Kerr, state geologist, in preparation of map of North Carolina. Asst. engr., Western North Carolina RR Co., 1867. Asst. engr. on line from Greensboro to Salem NC, 1868. Asst. engr., then chief of party on surveys and locations, also resident engr., Raleigh and Augusta Air Line RR Co., Cape Fear and Yadkin Valley RR Co. and Wilmington, Columbia and Augusta RR Co., 1868-1874. Prof. of mathematics and engr. at Carolina Military Inst., Charlotte, NC, Sept. 1874-Feb. 15, 1880. Chief of party on these projects from 1880-1882: lines from Lenoir, NC via Cook's Gap on the Blue Ridge to Elizabethton, TN; the Midland RR from Salisbury via Asheboro, to Goldsboro, NC; and the Atlantic Coast Line RR from Wilson, NC via Fayetteville, to Pedee, SC. Elected prof. of mathematics and engr. at the Military College of South Carolina, Charleston, Oct. 1882-Sept. 1888. Prof. of mathematics and engr. at the Univ. of North Carolina, Chapel Hill, 1888-1920. Upon retirement became Professor Emeritus. One of original five Kenan professors chosen as a mark of distinction by faculty of the Univ. of North Carolina in 1918. Student Chapter of ASCE named "William Cain Civ. Engr. Society" in his honor. Hon. degrees of LL.D. from the Univ. of South Carolina (1916) and D.S. from the Univ. of North Carolina (1920). Mem: ASCE (dir., 1912-1914), Am. Mathematical Soc., Am. Assn. for the Advancement of Science (fellow). Publ: "Uniform Cross-Section and T Abutments" (1874), *A Practical Theory of Voissoir Arches* (1874), "Trusses With Superfluous Members," "Positions of Live Loads Giving Maximum Strains for Single-Intersection Trusses," *Maximum Stresses in Framed Bridges* (1878), *Voussoir Arches, Applied to Stone Bridges, Tunnels, Domes and Groined Arches* (1879), republished in 1902 as *Theory of Steel-Concrete Arches and of Vaulted Structures*. Also *Theory of Solid and Braced Elastic Arches, Symbolic Algebra and Notes on Geometry* (1884), *Practical Designing of Retaining Walls* (1888), "Unit Stresses," "Determination of the Stresses in Elastic Systems by the Method of Least Work," *Maximum Stresses in Framed Bridges* (1897), "Stresses in Masonry Dams," "Experiments on Retaining Walls and pressures on Tunnels," "A Shortened Method in Arch Computation," *A Brief Course in the Calculus* (1905), *Earth Pressure, Retaining Walls and Bins* (1916), "The Circular Arch Under Normal Loads" (J. James R. Croes Medal, ASCE, 1923), and numerous other papers on long columns, high masonry dams, and portal bracing. Refs: TASCE, WWW, DAB, WAB, WWE.

CAMPBELL, ALLAN; b. Albany, NY, Oct. 11, 1815; d. NYC, March 18, 1894; f. Archibald; m. Margaret (Adams); w. Julia Farlie Copper; c. 2 daughters and 2 sons, including Col. John Campbell. Educated at the Albany Academy. Engr. of the Ithaca & Oswego RR, 1832-36. Civ. engr. Erie Canal and Ohio River improvement, 1836-50. Served for several years as engr. in the const. of the earliest RRs in the state of GA. Left the US in 1850 to start survey and const. of Copiapo to Caldera route, Chile, first RR operated in South America. Returned to US in 1856, became chief engr. of the extension of the Harlem RR and then pres., 1856-62. Pres. of the Consolidation Coal Co. During Civil War worked on harbor defenses of NYC. Chief engr. of const. of Union Pacific RR. Appt. commissioner of public works in NY, Jan. 21, 1876. Accepted comptrollership of NYC, 1880. Also appt. in the same year assessment commissioner of NY. Mentioned as a candidate for the governorship of NY, 1882. Nominated as a non-partisan candidate for mayor of NYC, defeated 1882. Mem: ASCE (hon. mem.), Geographical Soc., NY Hist. Soc., St. Andrews Soc., Century Club. Refs: DAB, WWW, Lamb, WAB, BDNA.

CARLL, JOHN FRANKLIN; b. Bushwick (Brooklyn), NY, May 7, 1828; d. Waldron, AR, March 13, 1904; f. John; m. Margaret (Walters); w. Hannah A. Burtis, 2nd w. Martha Tappan. Educated at Union Hall Academy in Flushing, NY. After grad. in 1846, farmed for three years. Entered publishing field and assisted in editing and printing of the *Daily Eagle* in Newark, NJ, 1849-53. Returned to Flushing in 1853, practiced civ. engr. and surveying for ten years. Moved to Pleasantville, PA in Oct. 1864 and worked in development of oil fields. Invented static-pressure sand pump, a removable pump chamber and adjustable sleeves for piston rods used in operation of oil wells. Involved in geological survey of PA in 1874, appt. asst. geologist in charge of petroleum and natural gas surveys. First geologist to comprehend structure of the oil regions of PA. Continued with survey until 1885. Resigned to enter private practice as consulting geologist. Mem: Pennsylvania Geological Soc. Wrote papers describing petroleum deposits to the annual reports of 1874-85 known as I, 1874; II, 1877; III, 1880; IV, 1883; and V, 1885. Refs: Lamb, WWW, DAB, CAB, WAB.

CASEY, THOMAS LINCOLN; b. Madison Barracks, NY, May 10, 1831: d. Washington, DC, March 25, 1896; f. Gen. Silas; m. Perry (Pearce); w. Emma Wier; c. Thomas Lincoln and Edward Pierce. Appt. to US Military Academy, July 1, 1848. Grad. July 1, 1852 at head of his class. Appt. brevet 2nd lt. in the Corps of Engrs. Promoted 2nd lt. June 22, 1854. Asst. engr. on harbor works of Delaware Bay and River and const. of Ft. Delaware until 1854. Asst. prof. of civ. and military engr. at US Military Academy 1854-1859. Reached rank of 1st lt., Dec. 1, 1856. In command of engr. soldiers on Puget Sound, Washington Territory 1859-61. Appt. Captain of Engrs. Aug. 6, 1861. Served during Civil War as engr. on staff of general commanding dept. at Fortress Monroe, VA. Supt. engr. in const. of forts and batteries on coast of ME. Special duty with North Atlantic Squadron during 1st expedition to Ft. Fisher, NC, Dec. 1864. Major of engrs., Oct. 2, 1863. Brevetted lt.-col. and col., March 13, 1865. Took seven month leave to put Portland Co. into efficient operation, July 26, 1866-

Feb. 25, 1867. In charge of division of fortifications in War Dept., Washington, DC, Nov. 18, 1867. Sent to Europe, summer 1873 at head of bd. to examine systems of torpedo const. adopted in Great Britain, Germany, Austria and France. Lt. col. of engrs., Sept, 1874. Given charge of public buildings and grounds in the District of Columbia and const. of buildings for the state War and Navy Depts., March 3, 1877-May 1888. Engr. of Washington aqueduct and directed improvements over grave of Thomas Jefferson at Monticello, VA. Selected as engr. and architect to complete Washington National Monument, June 25, 1878-Dec. 6, 1884. Member of bd. to advise on ventilation of hall of House of Representatives in the Capitol, 1877-86. Promoted col., Corps of Engrs., March 13, 1884. Pres., Bd. of Engrs. in NYC on July 1, 1886-1888. Mem. of Lighthouse Bd., 1884-1892. Appt. brig. gen. and Chief of Engrs., July 6, 1888. Designated to erect new bldg. for Library of Congress, Oct. 2, 1888. Appt. to commission to locate large park in suburbs of Washington, DC, Sept. 1890. Retired May 10, 1895. Mem: Soc. of the Cincinnati of Massachusetts, New England Hist. Soc., National Academy of Sciences, Legion of Honor of France. Refs: Lamb, DAB, WAB, BDNA.

CASS, GEORGE WASHINGTON, JR.; b. near Dresden, OH, March 12, 1810; d. NYC, May 21, 1888; f. George Washington; m. Sophia (Lord); m. 2nd Ellen Dawson; c. five sons and seven daughters. Attended Detroit Academy, 1824. Appt. from Ohio to the US Military Academy, 1827, grad. 1832. Duty with Corps of Topographical Engrs. in survey of Provincetown Harbor, MA, Sept. 12-Dec. 5, 1832. Became 2nd lt. on March 4, 1833, 1st lt., Dec. 3, 1865. Asst. to supt. in charge const., Cumberland Road, U. S. Army Corps of Engrs., 1832-Oct. 26, 1836. Continued in the service of the Corps as a civ. engr. until 1840. Erected first cast iron bridge (with Richard Delafield) built in the US, over Dunlap's Creek, a tributary of Monongahela River, 1837. Est. a mercantile business in Brownsville, PA in 1840. One of the engrs. in charge of the improvement of the Monongahela River. Financing by the state of PA fell through, instrumental in forming a private co. which completed the work, 1844. Est. Adams Express Co., 1849. Effected consolidation of company lines between Boston and St. Louis and south to Richmond, VA 1854, pres. of the co., 1856-62. Dir., Ohio & Pennsylvania RR Co., July 31, 1856-May 25, 1881. Continued as dir. until his death. Member of the Smith Syndicate which took possession of the franchise and debts of the Northern Pacific RR Co., 1866. Pres. of Northern Pacific RR Co., 1872-1875. Democratic candidate for governor of PA. Mem: Bd. of Visitors to the US Military Academy. Refs: DAB, WWW, WAB, TASCE.

CASSATT, ALEXANDER JOHNSTON; b. Pittsburgh, PA, Dec. 8, 1839; d. Dec. 28, 1906; f. Robert S.; m. Katherine Kelso (Johnston); w. Lois Buchanan; c. Capt. E. B., Robert K., Mrs. James Hutchinson and Mrs. W. Plunkett Stewart. Educated in the public schools of Pittsburgh. Went with family to Europe and attended schools in Paris, France, Heidelberg, Germany and there at Darmstadt Univ., 1851-56. Entered Rennsselaer Polytechnic Inst., Jan. 1857, grad. as civ. engr. in 1859. Engr. asst. on the Dalton & Jacksonville RR, GA. Returned to Philadelphia on outbreak of Civil War. Entered service of the Pennsylvania RR as rodman of the Philadelphia Division, 1861. Asst. engr. on the connecting line linking the Pennsylvania to the Philadelphia & Trenton RR, 1863. Resident engr. of the Middle Division in 1864. Gen. supt. of the Pennsylvania RR, April

1, 1870. Gen. mgr. of all Pennsylvania lines east of Pittsburgh and Erie, Dec. 1871. Elected third vice pres. in charge of transportation and traffic, 1874. 1st vice pres., 1880. Retired from active duty, Sept. 30, 1882. Re-elected a dir. of the co., Sept. 12, 1883. Appt. chair of the road committee. Pres. of the NY, Philadelphia & Norfolk RR, 1885-June 11, 1899. Spring, 1891, appt. US Representative of the Intl. Railway Commission, chosen pres. Pres. of the Pennsylvania RR, June 1899-1906. Built tunnel under the Hudson River, Manhattan Island and the East River, NY to connect with the Long Island RR. Refs: WAB, WWW, DAB, RPI Bio, CAB, CCI.

CHITTENDEN, HIRAM MARTIN; b. Yorkshire, NY, Oct. 25, 1858; d. Seattle, WA, Oct. 9, 1917; f. William Fletcher; m. Mary Jane (Wheeler); w. Nettie M. Parker; c. Eleanor Mary, Hiram Martin Jr., Theodore Parker. Student at Cornell for six months, appt. to US Military Academy, grad. with high honors as 2nd lt., Corps of Engrs., June 15, 1884. Entered the Engr. School of Application for three year course. Made 1st lt., Dec. 31, 1886, grad. 1887. Ordered to Omaha as engr. officer of Dept. of the Platte. Held appointment for two years, prepared topographical map of Colorado, Wyoming, Utah and portions of nearby states. Assigned to improvement of Missouri River above Sioux City, IA, 1889-1891. Asst. to officer in charge of road const. in Yellowstone National Park, June 1891- March 1893. Assigned to duty on the Louisville and Portland Canal, 1893. Fall 1894, executive officer of bd. of engrs., had charge of canal survey between Lake Erie and the Ohio River. Made capt. Oct. 2, 1895. Secy. of Missouri River Commission in charge of improvement of Osage and Gasconade Rivers in Missouri, surveys on Missouri River and surveys for reservoir sites in Wyoming and Colorado, 1896-97. Served as lt.-col. of Volunteers and chief engr. of 4th Army Corps in Spanish-American War. In charge of road const. in Yellowstone, spring 1899-1906. Promoted to rank of maj., Jan. 23, 1904. Placed in charge of Lake Washington Canal project to connect Puget Sound with lakes in and bordering the city of Seattle, 1906-16. Appt. chair of Federal Commission on Yosemite Park to consider changes in boundaries. Engaged in engr. projects of the Pacific Coast 1906-1908, including commission of engrs. to investigate Sacramento Flood Control. Lt.-col, 1908. Retired with rank of brig. gen. as a result of stroke causing partial paralysis, Feb. 10, 1910. Pres. of Port Commission of Seattle, planning and const. of docking and terminal facilities, Sept. 5, 1911-Oct. 15, 1915. Consultant for flood control to Spring Valley Water Co. of San Francisco, CA, 1912. Also reported to Miami Conservancy District on that problem. Mem: ASCE, Pacific Northwest Soc. of Engrs., Rainier and Arctic Clubs, Seattle, Missouri Hist. Soc. (hon. mem.), Phi Beta Kappa (charter and hon. mem.). Publ: *The Yellowstone National Park, Historical and Descriptive* (1895), two technical works on reservoir systems in 1897 and 1898, *The American Fur Trade of the Far West* (1902), *The History of Early Steamboat Navigation on the Missouri River* (1903), *The Life, Letters and Travels of Father Pierre Jean de Smet* (in collaboration with Alfred T. Richardson, 1905), *War or Peace* (1911), *Flood Control* (1915), *Letters to an Ultra-Pacifist* (1916). Refs: TASCE, WWW, DAB, WAB.

CHURCH, GEORGE EARL; b. New Bedford, MA, Dec. 7, 1835; d. London, England, Jan. 5, 1910; f. George Washington; m. Margaret (Fisher); w. Alice Helena Carter; w. 2nd Anna Narion Chapman; c. none. Moved to Rhode Island, 1843. Attended Providence High School. Asst. topographical engr. on the state map of MA, 1853-54. Asst. engr. on Mississippi and Iowa Central RR, 1855. Appt. resident engr. on the Hoosac Tunnel, 1856. Chief engr. on location of a long line of RR in IA. Went to Buenos Aires, Argentina as chief engr. of a RR project, 1857. Work postponed because of disturbances in the country. Appt. by Argentine govt. as member of scientific commission to explore the southwestern frontier and report on best system of defense against various tribes in the region. Surveyed and located the Great Northern RR of Buenos Aires, 1860. Chief asst. engr. during its const., 1860-61. Returned to the US at the start of the Civil War, commissioned as capt. in the 7th Rhode Island Infantry, July 27, 1862. Served in several capacities during the war including lt. col. of the 7th Regiment, Jan. 7, 1863, col. of the 11th Regiment, Feb. 11, 1863 and 2nd Regiment Rhode Island Volunteers, Dec. 30, 1864, and as a brig. com. in the Army of the Potomac. At the end of the war was appt. chief engr. of the Providence, Warren and Fall River RR. Served in Mexico, planned campaign resulting in the capture of Maximilian, 1866-67. Acted also as war correspondent and editor on the *New York Herald*, 1868. Went to South America, explored large part of the upper basin of the Amazon, 1868. Opened up new channels for trade between Bolivia and Brazil. Pres. of the National Bolivian Navigation Co. and chair of the Madeira and Mamore RR. Obtained concession to contruct RR to avoid the falls of Madeira River, financed by investors from Europe. Terminus of the RR eventually named in his honor. Appt. US commissioner to visit Ecuador and report on the country's political, financial and trade conditions, 1880. Engaged in enterprises in London, was sent to build RR in Argentina, 1889. Spent three months in Costa Rica representing foreign bondholders and made elaborate report to the Costa Rica RR Co. on the condition of their line. Represented ASCE at the Intl. Congress of Hygiene and Demography held in London. Mem: ASCE, Military Order of the Loyal Legion of the US, Council of the Hakluyt Soc., Royal Geographic Soc. (vice pres.), Royal Anthropological Inst. of Great Britain and Ireland. Publ: Report to the US, "Ecuador in 1881," "Argentine Geography and the Ancient Pampean Sea," *Aborigines of South America* (1912). Refs: TASCE, WAB, DAB.

CLYDE, GEORGE DEWEY, b. Springville, UT, July 21, 1898; d. Salt Lake City, UT, April 2, 1972; f. Hyrum Smith; M. Elanora Jane (Johnson); w. Ora Packard; c. Ruth, Ned, Richard, Jerald, Mary Ann. Grad. Springville, UT High School, 1917; grad., valedictorian, Utah Agricultural College, Logan, UT, 1921; B.S., agricultural engr., Phi Kappa Phi, Sigma Tau; M.S., civ. engr, Univ. of California, Berkeley, 1923; joined faculty Utah Agricultural College. Dean of School of Engr. and Technology, 1935-45. Concurrently, Dir., UT State Engr. Experimental Station, and dir., War Training Programs, 1940-45. 1945 to 1953, Chief, Division of Irrigation Engr. and Water Conservation, US Dept. of Agriculture, Soil Conservation Service. Strongly promoted development of new techniques, equipment and use of snow surveys to forecast stream flow in US. Authored over 50 technical publications on use of water. Dir., UT Water and Power Board 1953-56. Interstate Streams Commissioner for UT. Member, Upper Colorado River Commission, Bear River Commission, Columbia River Commission. Strongly advocated development of Colorado River water by promoting legislation authorizing water resource projects. As political novice, won Republican nomination for Governor of UT in 1956 over Republican incumbent. Won general election in 1956 over incumbent governor and Democratic candidate in 3-way election. Re-elected for second term 1960.

Reorganized several depts. of state government, replacing politically-appointed commissioners with professional directors, increased funding for state schools and highway const., planned state building program, initiated state library, started state parks system and used professional engr. expertise to protect state's right to Colorado River water. During second term helped shape final legislation to create Canyonlands National Park, in southeastern UT. Consulting engr., pres., Clyde, Criddle and Woodward, 1965-72. ASCE's Civil Government Award 1965, and Royce Tipton Award 1972. Mem: ASCE (pres., UT Section, 1940, fellow, hon. mem., 1963), Am. Soc. of Agricultural Engrs., Soil Conservation Soc. of Am., Western Snow Conference, National Reclamation Assn., Kiwanis, Rotary. High Priest, Church of Jesus Christ of Latter-Day Saints.

COGSWELL, WILLIAM BROWNE; b. Oswego, NY, Sept. 22, 1834; d. June 7, 1921; f. David; m. Mary (Barnes); w. Mary N. Johnson; w. 2nd Cora Louise Brown; c. Mabel. Family moved to Syracuse when he was four years old. Early education at Hamilton Academy and in private schools in Syracuse and Seneca Falls. Two years spent on engr. party on the survey of the Oswego & Syracuse RR. Entered RPI, May 1, 1850, left in 1852, hon. degree in civ. engr. in 1884. Apprenticeship in the Lawrence, MA machine shop, 1852-55. Mgr. of the machinery dept. of the Marietta & Cincinnati RR at Chilliocothe, OH, 1856-59. Supt. of the Broadway Foundry, St. Louis, MO, 1859-60. One founder of Sweet Bros, & Co. at Syracuse, NY. Appt. chief engr. in the US Navy, 1861. Fitted out five repair shops for different stations on the Atlantic seaboard and the Gulf of Mexico, commanded one of them erected on ship at Port Royal, SC. Transferred to the Brooklyn Navy Yard in charge of steam repairs, 1862-66. Resigned from naval service in 1866. The next two years lived in NYC. Went to Oswego, NY to build iron bridge over the Oswego River, 1868. Took charge of the Clifton Suspension Bridge at Niagara Falls, NY and at the same time employed at Franklin Iron Works in the const. of two blast furnaces, 1870-74. Took charge of the Mine La Motte lead mines, MO, 1874. Returned to Syracuse in 1879 to examine salt and brine deposits of the vicinity. Went to Europe to examine processes employed in soda manufactories there, planned with Solvay & Cie. of Brussels, Belgium to establish a branch in the US. Started the const. of the Solvay Process Co. works at Syracuse NY, 1881, became gen. mgr. and treas. Branch of the co. est. at Detroit, MI, 1887. Instrumental in the development of the Hanawa Falls Co., St. Lawrence Co. NY. Mem: ASCE, Am. Soc. of Mining Engrs. (vice pres.), Am. Soc. of Mech. Engrs. (mgr.), Geographic Soc., Soc. of Chemical Industry of London England, Am. Assn. for the Advancement of Science, NY Chamber of Commerce. Refs: WAB, RPI Bio, DAB, WWW.

COLLINGWOOD, FRANCIS; b. Elmira, NY, Jan. 10, 1834; d. Aug. 18, 1911; f. Francis; m. Elizabeth (Kline); w. Eliza W. (Bonnett). Attended academy in Elmira, entered RPI, spring 1853, grad. first in class, 1855, degree of civ. engr. Compass surveys near Elmira, NY, 1855-69. Occupied in RR engr. work in Wisconsin, 1857. Became city engr. of Elmira 1856-68, put in first permanent sewer system. Made preliminary surveys for RR in PA. Engaged in sale of scientific instruments and jewelry business, 1857-July 1869. Asst. engr. on Brooklyn Bridge, July 1869-July, 1883. Made extensive repairs to Allegheny Suspension Bridge at Pittsburgh, PA, 1883-84. Summer 1884 spent in Europe. Regular contributor to *Sanitary Engineer* from fall 1885. Opened office in NYC as consulting and expert engr. Chief engr. during const. of Newport News Dry Dock, 1887-89. On commission of Engrs. to examine work of new Croton Aqueduct, 1888-89. Expert examiner for Civ. Service Commission, NY, 1895. Lecturer on foundations, New York Univ., 1895-1904. Mem: Am. Inst. of Mining Engrs., ASCE (dir., 1873-76, secy., 1891-1894), Inst. of Civ. Engrs. of Great Britain, Rensselaer Engrs. Soc., Am. Geographical Soc., Metropolitan Museum of Art, NY Microscopical Soc., NY Academy of Science, Am. Assn. for the Advancement of Science, Am. Inst. of Architects, one founder of Elmira Academy of Science. Publ: several papers on East River Bridge, Allegheny Suspension Bridge (awarded Telford Premium and Telford Medal by Inst. of Civ. Engrs., 1884), wind pressures, preservation of forests, cement testing, tests of steel and iron, and power of water. Refs: WWW, Lamb, RPI Bio, TASCE, WAB, BDNA.

COOLEY, LYMAN EDGAR; b. Canandaigua, NY, Dec. 5, 1850; d. Feb. 3, 1917; f. Albert Blake; m. Achsah (Griswold); w. Lucena McMillan; c. 2 sons and 1 daughter. Taught in Canandaigua Academy, 1870-72. Entered RPI, 1872, grad. in 1874 with degree in civ. engr. Prof. in charge of engr. at Northwestern Univ., 1874-77. Assn. editor of the *Engineering News*, summer 1876-May 1878. Asst. engr. on const. of bridge over the Missouri River at Glasgow, MO, 1878. Asst. engr. on the Mississippi and Missouri River improvements and charge of local improvements and surveys in NE, IA, WI, AR and TN, 1879-84. Editor for one year of the *American Engineer*, 1884. Mem. of a sub-committee of the Citizens Assn., drew report in Sept. 1885 in favor of a sanitary canal and secured organization of drainage and water supply commission. Chief asst. of that commission in 1886-87. Consulting engr. to Chicago and the Sanitary Commission. Engr. to commission that determined boundaries of the Sanitary District in 1889. Mem., Bd. of Trustees and chair, engr. committee, autumn 1890-Dec. 1895. Consulting engr. of the Sanitary District in 1897. Worked on solution to problems regarding the Chicago Sanitary District including its relation to lakes and rivers, sanitary and const. features. Appt. by Pres. Cleveland to Intl. Deep Waterways Commission to make investigations for a ship canal between the Great Lakes and the Atlantic Ocean, 1895-96. Went to Panama and Nicaragua as consulting engr. with the contractors and engrs. expedition regarding the Isthmian canal problem, 1897-98. Advisory engr. in investigation for the Erie Canal, 1898. Consultant for the Denver Union Water Co. on the Cheesman Dam, 1899-1904. Worked in connection with water power enterprises and water storage at Keokuk, IA. Mem. of the US Postal Committee on Pneumatic Tubes for mail in cities, 1901. Advising engr. on water works appraisement, Omaha, NE, 1904. Consulting engr. on location of barge canal, Rochester, NY, 1905, in Grand Rapids, MI on flood problem, 1905-1910, and in Saginaw, 1912. Secy. and consulting engr. of the Intl. Improvement commission of Illinois, 1906-1909. Consulting engr. for the Lakes-to-the-Gulf Deep Waterway Assn., 1909. Chair of Commission on Sewage Disposal and Water Power Development, Sanitary District of Chicago, 1912-15. Mem: ASCE, Chicago Academy of Sciences, Rensselaer Soc. of Engrs., Soc. of Western Engrs., Am. Water Works Assn. Publ: *The Diversion of the Waters of the Great Lakes by way of the Sanitary and Ship Canal of Chicago* (1913), *Isthmian Canal* (1902), also many papers in the technical press. Refs: DAB, WWW, RPI Bio, WAB.

CRANDALL, CHARLES LEE; b. near Bridgewater, NY, July 20, 1850; d. Aug. 25, 1917; f. Peter B.; m. Eunice C. (Priest); w. Myra G. Robbins. Educated in district schools, 2 years at an academy. Family moved to Ithaca, NY, 1868. Entered Cornell, Oct. 8, 1868. Grad. 1872, degree of civ. engr. Served in architect's office and as asst. RR engr. on NY, Boston & Montreal RR, 1872-1874. Entered Cornell again as a grad. student, Jan. 1874. Instr. of civ. engr. April 1874, asst. prof., July 1875. Received master of civ. engr. degree, 1876. Assoc. prof., June 1891, prof. RR engr. and geodosy, June 1895, prof. RR engr., 1908-15. Aid to US Coast Survey, summer, 1878. City engr., Ithaca, NY 1879-91. Served on Committee on Iron and Steel Structures of Am. Railway Engr. Assn., 1901. Prof. in charge of College of civ. engr., Cornell, 1902-1906. Made Prof. Emeritus, 1915. Served on Sewer Commission and served term as Commissioner of Public Works in Ithaca, 1917. Mem: ASCE, Sigma Xi, Tau Beta Pi, Civ. Engr. Soc. Semaphore, Zodiac Fraternity, Soc. for the Promotion of Engr. Education (pres., 1906), Am. Railway Engr. Assn. Publ: *Tables for Computation of Railway and Other Earthwork* (1886, 1893, 1902, 1907), *Notes on Descriptive Geometry* (1888, 1893), *Notes on Shades, Shadows and Perspective: The Transition Curve* (1893, 1899), *Textbook on Geodesy and Least Squares* (1907), *Field Book for Railroad Surveying* (1909), *Field Book for Railroad Construction* (1913), and several contributions to *Van Nostrand's Engineering Magazine* and the *Transactions* of ASCE. Refs: TASCE, Lamb, WWW, WAB, BDNA.

CURTIS, SAMUEL RYAN; b. Champlain, NY, Feb. 3, 1805; d. Council Bluffs, IA, Dec. 26, 1866; f. Zarah; m. Phalley (Yale); w. Belinda Buckingham. Grad. US Military Academy, 1831. Assigned to the 7th Infantry and sent to Ft. Gibson. Resigned his commission in 1832 and returned to OH. Engr. employed on the National Road. Chief engr. of the Muskingum River improvement project, April 1837-May 1839. Studied law and had law office at Wooster, OH. Adjunct gen. then col. 3rd Ohio Infantry during the Mexican War. Chief engr. of improvement, Des Moines River in Keokuk, IA until Dec, 1849. City engr. St. Louis, 1850-53. Const. dike to deepen the channel on the St. Louis side of the Mississippi River. Promoted the American Central RR, 1853. Opened law office in Keokuk, IA, 1855. Elected mayor, 1856. Member of the US House of Representatives from IA, 35-38th Congress, 1857-62. Chair of a special congressional committee to make plans for the Pacific RR. At start of Civil War was chosen col. of the 2nd Iowa Infantry sent to protect the RRs in MO. Appt. brig. gen. US Army at the special session of Congress, July 4, 1861. In command of the southwest district of MO, Dec. 1861-Feb. 1862. Commanded the Army of the Southwest, Feb.-Aug. 1862. Defeated Confederates at Pea Ridge, AR, spring 1862. Maj. gen., March 21, 1862. In command of the Dept. of Missouri, 1862-63. Assigned to Dept. of Kansas, 1864-65, and Dept. of Northwest, spring 1865. Peace Commissioner to Indians, Aug.-Nov. 1865. Mem. of commission to inspect Union Pacific RR, Nov. 1865-April 1866. Retired from the Volunteer Service, April 30, 1866. Refs: WWW, DAB, WAB, BDNA, DR, CAB.

DELACY, WALTER WASHINGTON; b. Petersburg, VA, Feb. 22, 1819; d. Helena, MT, May 13, 1892; f. William De Lacy; m. Eliza (Lee); not married. Entered Mt. St. Mary's Catholic College, MD, 1834, grad. 1838. Studied under Prof. Mahan of US Military Academy. Const. engr. for the Illinois Central and Mountain RRs. Asst. prof. of French at US Military Academy, 1840. Prof. of languages for five years at the US Naval Academy. Searched for abandoned mines in Texas, 1846. Volunteered at the start of the Mexican War and served throughout. Engaged in RR work afterward, helped survey the 32nd parallel from San Antonio to San Diego. Served in the Nez Perce War, 1855. Asst. in building the Mullan Road from Ft. Benton, MT to Walla Walla, WA, 1858. In charge of const. from Sohon Pass to the Bitterroot River. Made map and report of the Bitterroot country, 1860. Prospector for gold in MT, 1861. Led gold seeking party into the Snake River country and discovered Shoshone Lake and the Lower Geyser Basin of the Yellowstone Park, 1863. Employed by the first legislature of MT in making the first state map, 1864. Located the initial point for public surveys in MT. Surveyed and located the line of the Northern Pacific. Employed in the surveyor general's office in Helena, MT, 1868-71, 1885-92. City engr. of Helena, 1883-84. Mem: Montana Soc. of Engrs. (founder), Montana Hist. Soc. Refs: DAB, WAB, WWW.

DELAFIELD, RICHARD; b. NYC, Sept. 1, 1798; d. Washington, DC, Nov. 5, 1873; f. John; m. Ann (Hallett); w. Helen Summers; w. 2nd. Harriet Covington; c. 5 daughters and 1 son. Grad. US Military Academy, first in class, 1818. Immediately promoted 2nd lt. of engrs. Assigned to duty on Northern Frontier in boundary commission est. under treaty of Ghent. In 1819-24 served as asst. engr. in the const. of Forts Monroe and Calhoun, next assigned to Mississippi River, taking charge of defenses of Plaquemine Bend, surveys of the Delta and general supervision of improvements. Supt. engr. in const. of the Cumberland Road, in building Fort Delaware, in repairing Fort Mifflin and in improvement of the Delaware River harbors and breakwater. Promoted 1st lt. in 1820, capt., May 24, 1828, maj., July 7, 1838. Designer of 80-ft.span iron bridge constructed at Dunlap's Creek, Brownsville, PA, 1836-39 (first cast iron bridge in the U.S). Supt., US Military Academy, Sept, 1838-Aug. 1845, Sept. 8, 1856-March 1, 1861. Supt. engr., NY Harbor Defenses, NY Lighthouse District, and Hudson River improvements, 1845-56. During 10 months of that time, was chief engr. of the Dept. of Texas. Accompanied Capt. George B. McClellan and Maj. Alfred Mordecai to Europe to witness operations of the Crimean War, 1855-56. Submitted extensive report on the art of war in Europe during 1854-56. In 1856 appt. supt. of the US Military Academy. Relieved of post at own request, April 1861. Promoted to lt. col. in 1861, col. in 1863. Served on staff of Gov. Morgan of NY in reorganization and equipment of state forces, 1861-63. During that same time, supt. const. of NY harbor fortifications. Brig. gen., Chief of Engrs. US Army, April 22, 1864. In command Corps of Engrs., Inspector of the US Military Academy; also in charge of Engr. Bureau, Washington, DC, 1864-66. Brevetted major gen., May 13, 1865. Mem. of the Lighthouse Bd., 1864-70. Retired Aug. 8, 1866. Mem: Commission for Improvement of Boston Harbor; regent, Smithsonian Inst. Refs: DAB, CAB, Lamb, WWW, WAB, CCI, DR, BDNA.

DUNLAP'S CREEK BRIDGE AT BROWNSVILLE, PA.
DESIGNED BY RICHARD DELAFIELD. CONSTRUCTED 1836-1839

DETMOLD, CHRISTIAN EDWARD; b. Hanover, Germany, Feb. 2, 1910; d. NYC, July 2, 1887; f. Johann; w. Phoebe Crary; c. Wilhelmina Emilie and Zella Trelawney. Educated in a military academy in Hanover. Came to the US on his way to Brazil to enter the army there, 1826. Instability in Brazil caused him to remain in the US. Surveyor in Charlestown, SC. Completed surveys for RR for Charlestown & Hamburg RR & Canal Co., 1830. Won $500 for designing the best locomotive used by this company. Worked for US War Dept., supervised const. of Ft. Sumter, SC, 1833-34. Surveyor on various Eastern RRs, 1834-44. Iron manufacturer, MD, 1845-52. Const. engr. of the Crystal Palace, World's Fair "Exhibition of the Industry of All Nations," 1851. Traveled in Europe for a few years, returned to NY in 1865. Publ: *The Historical, Political and Diplomatic Writings of Niccolo Machiavelli, Translation from the Italian* (1882). Refs: WWW, DAB, ASM, CAB.

DEVEREAUX, JOHN HENRY; b. Boston, MA, April 5, 1832; d. Cleveland, OH, March 17, 1886; f. John; m. Matilda (Burton); w. Antoinette Kelsey; Attended an academy in Portsmouth, NH. Moved to Cleveland, 1848. Apprentice engr. on various RRs, including the Cleveland, Columbus & Cincinnati RR Co. and the Cleveland, Painesville & Ashtabula RR. Worked as RR engr. on various roads. Enlisted during the Civil War, supt. of military RRs in Virginia, 1862-64. First supt. then vice pres., Cleveland & Pittsburgh RR, 1864-68. Vice pres. and gen. mgr., Lake Shore & Michigan Southern RR, 1868-73. Pres. of the Big Four RR (Cleveland, Columbus, Cincinnati and Indianapolis), 1873-86. Receiver of Atlantic & Great Western RR, 1874-80, pres., 1881. Agent for William K. Vanderbilt in negotiations for acquisition of the NY Chicago & St. Louis RR by the NY Central. Later advocated an interstate commerce commission. Mem: Am. Assn. for the Advancement of Science, Loyal Legion, and Cleveland Humane Soc. Refs: WWW, DAB, CAB, WAB.

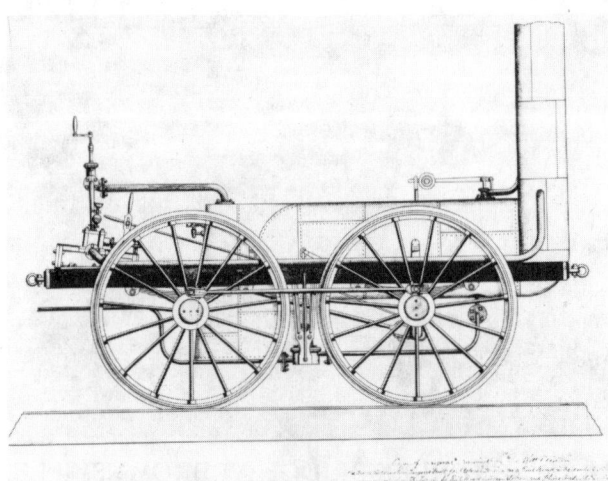

THE "DE WITT CLINTON" LOCOMOTIVE. BUILT IN 1831 FOR THE MOHAWK & HUDSON RAILROAD BETWEEN ALBANY AND SCHENECTEDY, NY

DILLON, SIDNEY; b. North Hampton, Montgomery County, NY, May 12, 1812; d. NYC, June 9, 1892; f. Timothy; w. Hannah Smith; c. Julia and Cora. Began as water carrier to workers on the Mohawk & Hudson RR from Albany to Schenectady in 1830-33, next employed in same capacity on the Rensselaer & Saratoga RR. Overseer for Crane & Clark, who had contract on the Boston and Providence RR. Foreman for the same firm on the Stonington Road in CT and on heavy rock work near Charlton, MA, on the Boston & Albany RR, then the Western RR. Won bid for const. of small section of the Boston & Albany RR near Hinsdale, completed 1840. Next contract on the Troy and Schenectady RR. Member of the firm Boody, Ross & Dillon which built a portion of the Cheshire RR in VT and the Hartford and Springfield RR. Engaged in const. of the Rutland and Burlington RR, Central RR of NJ, Philadelphia & Erie RR, Morris & Essex RR, Pennsylvania RR, New Orleans, Mobile & Chattanooga RR, Canadian Southern RR. Also built tunnel from Grand Central Station to Harlem River, NYC. Const. Union Pacific RR, 1865-69, dir., 1864-92, pres., March 11, 1874- June 19, 1884, Nov. 26, 1890-April 27, 1892. Also served as dir. in the Western Union Telegraph Co., Manhattan Elevated RR Co., Missouri Pacific RR Co., Pacific Mail Steamship, Chicago, Rock Island and Pacific RR, Wabash RR, Canadian Southern RR and Wheeling and Lake Erie RR Co. Mem: ASCE. Refs: WWW, DAB, TASCE, BDNA, ASM, Lamb.

DONOVAN, JOHN JOSEPH; b. Rumney, NH, Sept. 8, 1858; d. Bellingham, WA, Jan. 9, 1937; f. Patrick; m. Julia (O'Sullivan); w. Clara Elizabeth Nichols; c. Helen Elizabeth, John Nichols, Philip Laurence. Educated in the public schools. Grad. Plymouth, NH State Normal School, 1877. Taught in schools in MA and NH, 1877-80. Studied at the Worcester Polytechnic Inst., grad. with degree in civ. engr. in 1882. Engr. on lines in MT and WA, Northern Pacific RR, 1882-88. Started as rodman on a surveying crew, within six months was asst. engr. on const. In charge of the western half of the Cascade Division, June 1887. Moved to Fair Haven (later Bellingham), WA, 1888. Chief engr., Fairhaven & Southwestern RR and allied co., 1888-91. Chief engr. and mgr., Blue Canyon Coal Mining Co., 1891-1902 and Bellingham Bay & Eastern RR CO., 1898-1906. Chief engr. and mgr., Bellingham Bay & British Columbia RR and allied co., 1906-1910. Helped to bring about extensions and consolidations and served as chief engr. for the other two transcontinental RRs, the Great Northern and the Chicago, Milwaukee & St. Paul to bring lines into Bellingham. Vice pres. of the First National Bank of Bellingham and of the mem. of the first State Highway Commission and pres. of the Washington Good Roads Assn., 1924. Chair of the State Commission on Forest Legislation. D.Sc. from Worcester Polytechnic Inst., 1932. Mem: ASCE, Montana Soc. of Engrs., Am. Irish Hist. Soc., National Geographic Soc., State Bd. of Charities and Correction, advisor to St. Joseph's Hospital, National Municipal League, National Child Labor Committee, Knights of Columbus, Bellingham Chamber of Commerce (pres.), Washington State Hist. Soc., Washington Art Assn., Am. Hist. Soc. and Navy League. Refs: WWW, WWE, WAB, DAB.

DOUGLAS, WALTER JULES; b. Baltimore, MD, Oct. 26, 1872; d. NYC, July 2, 1941; f. August; m. Mary Virginia (Crawford); w. Elizabeth Appleton Spaulding; c. Mary Virginia, Walter Spaulding, Elizabeth Appleton, Caroline Hewetson. Moved to Washington, DC in 1876. Educated in public and private schools. Attended Bethlehem Preparatory School for Engrs., Lehigh Univ., grad.

B.S. in civ. engr., 1894. Draftsman for Coxe Iron Manufacturing Co. at Drifton, PA engaged in manufacture of equipment for coal mines. Studied coal-breaker const. and the testing of anthracite coals. Transitman and asst. engr. for the Lehigh Valley Coal Co. Chief engr. of F. C. Dinney & Co., a VA mining co., 1898. Asst. engr. of bridges and highways for the District of Columbia, 1898. Engr. of bridges, 1903-1909. Designed and supervised const. of all bridges built within the city of Washington, DC, including the Connecticut Ave. Bridge (later the Taft Bridge), completed in 1906, the largest concrete bridge built in the US. Also built the Anacositia and Piney Branch Bridge, the first unreinforced parabolic concrete arch. Assoc. with the firm of Marclay, Parsons & Klapp, 1909. In charge of all the structural design work of this firm and in this capacity collaborated on the design of the Almendares Bridge in Havana, Cuba. Paper on this bridge was awarded the Thomas Fitch Rowland Prize, ASCE, in 1912. Participated in design of the Mohawk hydroelectric development near Ephratah, NY and the Salmon River development near Altmar, NY. Designed and supervised the const. of a RR car manufacturing plant for the National Steel Car Co. at Hamilton, Ontario, Canada, 1911. Deputy chief engr. of the Cape Cod Canal Co., 1915-17. Participated in the organization of plants at Niagara Falls, Plattsburgh and Ft. Plain, NY for the manufacture of ferro-silicon during the early months of World War I. Engr. of maintenance of the Panama Canal and also vice pres. of the Panama Steamship Line and Panama RR Co., June 1918-May 1919. Acting governor for two months in the absence of the governor of the canal zone. Mem. of the firm Parsons, Klapp, Brinckerhoff & Douglass from 1919 until his death. Participated in const. projects in parts of the US, Canada, Cuba, Central America and Europe. Consulting engr. on the const. of the Arlington Memorial Bridge across the Potomac at Washington, DC, 1925-32. Supervised the design and const. of a vehicular tunnel and the design of a pedestrian tunnel under the Scheldt River at Antwerp, Belgium. Cmdr. of the Order of the Crown of Belgium, 1933. Consulting engr. on the US govt. tidal power project in Passamaquoddy, ME and chair, bd. of consultants on the Atlantic-Gulf Ship Canal to cross Florida. Consultant on foundations and structures of the NY World's Fair, on foundation problems at the La Guardia Airport and on the East River Drive, NYC. Various flood control projects in southern CA, the RR lift bridge over the Cape Cod Canal, MA, Jamestown Bridge over Narragansett Bay, RI, and the Pennsylvania Ave. Bridge over the Anacostia River, Washington, DC. Pres. of the Parklap Const. Co. and Parklap, Inc, which projects included the Sherman Island Dam and Power House, Feeder Dam Power House, RR and harbor tunnels at Detroit, Toronto, Montreal, Buffalo and Albany, wharfs and bulkheads at Detroit, Buffalo and Havana and certain sections of the Detroit-Windsor Vehicular Tunnel. Assoc. ed of Merriman's *American Civil Engineers Handbook* (1911). Mem: ASCE, Inst. of Civ. Engrs. of Great Britain, Soc. of Military Engrs., Am. Soc. for Testing Materials, Am. Railway Bridge and Building Assn., Washington (DC) Soc. of Civ. Engrs., Phi Delta Theta, Army and Navy,, Union and Engrs. Clubs of Washington, DC and the Plainfield Country Club. Publ: *Hints to Concrete Constructors* (1910), and many articles to scientific journals. Refs: WAB, YPB.

DOUGLASS, DAVID BATES; b. Pompton, NJ, March 21, 1790; d. Geneva, NY, Oct. 19, 1849; f. Nathaniel; m. Sarah (Bates); w. Ann Eliza Ellicott. Educated at home and by Rev. Samuel Whelpley. Grad. Yale, 1813. Commissioned 2nd lt. of engrs., Oct. 1, 1813. At US Military Academy, winter

1813. Summer 1814, head of corps of sappers and miners. Fought at battle of Lundy's Land. Const. and defended a battery, promoted 1st lt. and brevetted capt., 1814. Organization became known as the Douglass Battery. Promoted maj. Asst. prof. of natural and experimental philosophy in the US Military Academy, Jan. 1, 1815-20. Received A.M. from Yale and the College of New Jersey, 1819. Astronomical surveyor, US Boundary Commission, 1819-20. Prof. of mathematics at the US Military Academy, 1820-23. Prof. of civ. and military engr., 1823-31. Received A.M. from Union College, 1825. Resigned from the Army in 1831. In charge of const. of the Morris and Essex Canal, introduced inclined plane as a substitute for the lock system, 1831-32. Prof. of natural philosophy and civ. engr. at the Univ. of the City of New York, 1832-39. Engr. for commissioners of NYC, 1834-36. Plan for supplying NYC with water, 1835, adopted and resulted in the Croton Water Works, became 1st chief engr. Pres. and prof. of moral and intellectual philosophy, Kenyon College, 1840-45. Received LL.D. from Yale and Hobart College, 1841. Prof. of civ. engr. and architecture in the Univ. of the City of New York, 1844-53. Built the RR from Brooklyn to Jamaica, the supporting wall for Brooklyn Heights, planned the Greenwood Cemetery and introduced the water supply for Brooklyn. Laid out the Catholic Cemetery, Albany, NY, 1845-6. Engaged in developing the landscape features of Staten Island, NY, 1847, and in 1848 laid out the Protestant Cemetery at Quebec. Prof. of mathematics and natural philosophy in Geneva (later Hobart) College, 1848-49. Refs: BDNA, WWW, DR, DAB, Lamb, CAB.

DUANE, JAMES CHATHAM; b. Schenectady, NY, June 30, 1824; d. NYC, Nov. 8, 1897; f. James; m. Harriet (Constable); w. Harriet W. Brewerton; c. Alexander, James and another son. Entered Union College 1840, grad. 1844 with degree of A.B. Entered US Military academy, 1844, grad. July 1, 1848, 3rd in class of 38. Assigned as officer to the US Army Corps of Engrs. Served at US Military Academy with company of sappers, miners and pontoniers, 1848-54. Asst. instr. of practical military engr., 1852-54. Asst. engr. in building Ft. Trumbell, CT, 1849, and Ft. Carroll, MD, 1854-56. Engr. of NY Lighthouse District, 1856-58. On the Utah Expedition in command of the engr. co., 1858. Inspector of practical military engr. in command of the sappers, miners and pontoniers at the US Military Academy, 1856-61. Treasurer of US Military Academy, 1859-61. Commanded an engr. co. guarding the national capitol at Ft. Pickens, FL, 1861. Promoted capt. of engrs., Aug. 6, 1861. Organized the engr. batallion and engr. equipage, Army of the Potomac, winter 1861-62. Bridged the Potomac at Harper's Ferry, WV, Feb. 1862. Commanded engr. batallion in siege of Yorktown, VA, April 1862. Took part in the battle of Gaine's Mill, June 27, 1862. Built bridge 2000 ft. long over the Chickahoming, Aug. 12-14, 1862. Chief engr. of the Army of the Potomac, took part in the battles of South Mountain and Antietam. Promoted maj., March 3, 1863. Chief engr. of the Dept. of the South, 1863. Engaged in the attack on Ft. McAllister, GA and in operations against Charleston, SC. Chief engr., Army of the Potomac, July 15, 1863-June 1865. Brevetted lt. col. and col., July 6, 1864. Brevetted brig. gen., March 13, 1865. In command of the post of Willet's Point, NY, supt. engr. of the fort, 1865-68. Promoted lt. col on March 7, 1867. Promoted col., June 10, 1883. Engr. in charge of 1st, 2nd, 3rd Lighthouse Districts. Pres., Bd. of Engrs., NYC. Made chief of Engrs. with rank of brig. gen., Oct. 4, 1886. Retired June 30, 1888. Appt. mem. of the Bd. of Croton Aqueduct Commissioners of NYC, elected pres., Aug. 1888. Mem: ASCE (hon. mem.). Publ: *Manual for Engineer Troops, Organization of the Bridge Equipage of the US Army* (1870). Refs: Lamb, WWW, DAB, CAB, TASCE, BDNA, WAB.

DURFEE, WILLIAM FRANKLIN; b. New Bedford, MA, Nov. 15, 1833; d. Middletown, NY, Nov. 14, 1899; f. William; m. Alice (Talbot); w. Annie Swift. Practical mech. training under his father, special study in the Lawrence Scientific School at Harvard. Architect and civ. engr. in New Bedford, city surveyor for five years. Assoc. with his father in design and const. of Gosnold Iron Works, 1855. Representative in the state legislature, secy. of Committee on Militia, 1861. Designed a submerged gun for naval use, not adopted by govt. Designed and in charge of const. of experimental steel works at Wyandotte, MI, June 1862. Supervised the making of the first Bessemer steel produced in US, Sept. 1864. Rolled the first American steel rails, May 25, 1865. Also est. in connection with the plant at Wyandotte, the first steelworks analytical laboratory built in the US. Designed and const. rolling mill at Chicago machine shop for repair of small arms at Cambridge, MA. Built works of Milwaukee Iron Co. at Bayview, Wisconsin. Planned and built works of the American Silver Steel Co. at Bridgeport CT, 1869. Gen. mgr of William Butcher Steel Works, Philadelphia, PA. Supervised manufacture of steel used in St. Louis Bridge (Eads Bridge), first steel bridge erected in the US, 1871. Gen. supt. of the Milwaukee Iron Co. works at Bay View, WI, 1873. Mem., bd. of judges of Centennial Exhibition, awarded special medal, 1875. Designed special machinery for Wheeler and Wilson Manufacturing Co., Bridgeport, CT, 1878-82. Visited Europe, studied use of fuel gas in manufacture of copper and brass. Erected first gas furnace in US for refining copper. Engineered removal of a brick chimney 8 ft. square at its base, 100 ft. high, weighing 170 tons, a distance of 30 ft. to a new foundation for the Bridgeport Paper Co., 1885. Mgr. of US Mitis Co., 1886. Gen. mgr. of Pennsylvania Diamond Drill and Manufacturing Co. of Birdsboro, PA, erected large additions to the plant, 1887. Became connected with C. W. Hunt Manufacturing Co. of West New Brighton, NY as supt. after opening of co. as consulting mgr. and expert in patent cases. Mem: Am. Assn. for the Advancement of Science, Am. Inst. of Mining Engr., Am. Soc. of Mech. Engr. (mgr., 1883-6, vice pres., 1896-98), Franklin Inst., 1871, Iron and Steel Inst. of Great Britain, 1875, Assoc. of US Naval Inst., 1885. Refs: WAB, DAB, Lamb, BDNA, CAB, WWW.

DURHAM, CALEB WHEELER; b. Tunkhannock, PA, Feb. 6, 1848; d. Peekskill, NY, March 28, 1910; f. Alpha; m. Elizabeth B. Riggs; w. Clarissa Safford Welles; c. four sons. Moved to Reading, PA after the death of his father when he was four years old. Early education in PA. Enlisted during Civil War as a private, first in Co. C, 42nd Pennsylvania Militia, served in the reserves at Gettysburg and in Company B, 195th Pennsylvania Volunteers. Active service in MD and VA. Discharged in 1864. Worked for the Philadelphia & Reading RR Co. Attended Williston Academy, Easthampton, MA, 1866. Entered the Univ. of Michigan in 1867 to study civ. engr. Two years later gave up college and worked in the engr. dept. of the NY Central RR. Engaged in RR work for several cos. in the midwest and southwest, 1869-73. Sanitation engr. in Chicago, 1873. Devised an improved hot air heater and undertook its manufacture and sale, 1875. Abandoned when became too costly for use. Invented the Durham System for house drainage, 1880. Organized the Durham House Drainage Co.

Entire house drainage system for the city of Pullman, IL. Moved to NY, 1883 and reorganized his business there. Installed system in Carnegie Hall and the Hotel Majestic in NY, and the National Capitol in Washington, DC. Mem: Engrs. Club of the Northwest, Chicago, and the Engrs. Club, NY. Refs: DAB.

ENDICOTT, MORDECAI THOMAS; b. May's Landing, NJ, Nov. 26, 1844; d. March 5, 1926; f. Thomas Doughty; m. Ann (Pennington); w. Elizabeth Adams; c. Maud, Anna, Grace, Elizabeth, Edith, May and Louise. Elementary education through Presbyterian Church and private institutions. Entered RPI, Feb. 1965. Grad. in 1868, degree of civ. engr. Asst. in office of R.P. Rothwell, civ. and mining engr. at Wilkes-Barre, PA, July-Dec. 1868. Rodman, civ. engr. dept. of US Navy Yard, Brooklyn, NY, May 1869-Jan. 1870. Draftsman in office of New Haven, Middletown & Willimantic RR, NYC, Jan., 1870. Asst. on CT River bridge at Middletown, CT, May 1870. Asst. engr. in charge of the Dresden extension of Cincinatti & Muskingum Valley RR, Nov. 1870-Jan. 1872. Asst. civ. engr. at League Island Naval Station, Philadelphia, PA, Feb.-Oct. 1872. Asst. civ. engr. in charge of dept. of yards and docks at US Navy Yard, Philadelphia, PA. Commissioned as civ. engr. in US Navy, July 13, 1874. July 16, 1874, ordered to Navy Yard at New London, CT in charge of dept. of yards and docks. July 1879 ordered to Navy Yard at Portsmouth, NH. Ordered to Navy Yard at League Island, Philadelphia, PA, July, 1881. Remained in charge of const. of dry dock until completion in 1889. Given rank of Cmdr., 1882. Appt. mem. of bd. to consider and investigate conditions and needs at NY Navy Yard and report plan for development and improvement. Brought to Washington, April 1890, as consulting engr. in Bureau of Yards and Docks, US Navy Dept., in charge of all civ. engr. projects. Served on Nicaragua Canal Commission, 1895. Appt. by Pres. McKinley as chief of Bureau of Yards and Docks, April 7, 1898 until 1907. Mem., Armor Factory Bd. in 1897, and the Navy member of Panama Canal Commission, 1905-1907. Retired as rear-admiral in Nov. 1906, but remained on duty as technical advisor until June, 1909. Most notable achievement was completion of floating dry dock, *Dewey*, largest of its type up to that time. Returned to active duty during World War I, Oct. 12, 1917. Relieved from active duty June 30, 1920. Buried in Arlington National Cemetary. Mem: ASCE (pres., 1911), Franklin Inst., Washington Soc. of Engrs., Engrs. Club of NYC, Theta Xi Fraternity, New Jersey Hist. Soc., Cosmos Club, Monday Evening Club of Washington. Refs: TASCE, DAB, RPI Bio, WAB, WWE.

ERNST, OSWALD HERBERT; b. near Cincinnati, OH, June 27, 1842; d. Washington, DC, March 21, 1926; f. Andrew Henry; m. Sarah H. Otis; w. Elizabeth Amory; c. Helen and Mrs. William Morton Grinnell. Attended private schools, entered Harvard 1858-59. Received appt. to US Military Academy from the state of Ohio, entered July 1, 1860. Grad. near head of his class, June 13, 1864. Promoted to 1st lt., Corps of Engrs., US Army. Assigned to the Army of the Tennessee as asst. engr. Took part in battle and siege of Atlanta, July 22, 1864, and battle of Ezra Church, July 28, 1864. Received brevet of capt., March 13, 1865. Asst. engr. in const. of fortifications in harbor of San Francisco, CA 1864-68. Attained rank of capt. of engrs., March 1867. Instructor in Engr. School of Practice at Willit's Point, Long Island, 1868-1871.
Astronomer of US Commission sent to Spain to observe solar eclipse, Dec. 1870. Instr. in Practical engr. at the Military Academy, also architect of academy bldgs. erecting there, 1871-1878. From 1878-1886 in charge of river and harbor

improvements of Osage River, MO, and of Mississippi and Missouri Rivers at points in IL and MO. Attained rank of maj., May 1882. From 1886-1889 supervised digging of deep-sea channel to harbor of Galveston, TX. Supt. of public buildings and grounds at Washington, DC, 1889-93. Appt. col. of engrs., March 31, 1893. Supt. of US Military Academy 1893-98. Attained rank of lt. col. in 1895. Commissioned brig. gen. of Volunteers in Spanish-American War, May 26, 1898. Commanded 1st Brigade of the 1st Corps in Puerto Rican campaign, 1898 and 1899. Participated in engagements at Coamo and Asamante; received commendation from his superiors. Served as Inspector Gen. in Cuba 1898-99. Became mem. of Isthmian Canal Commission, visited Europe and Central America in connection with study of proposed route. Mem. of commission which determined that Panama Canal should have locks, 1905-1906. In charge of river and harbor improvements at Baltimore, 1900-1901; similar duty in Chicago, 1901-1905. Pres. of Mississippi River Commission 1903-1906. Chair, bd. of engrs. to survey route for 14-ft. deep waterway between Chicago, and St. Louis, 1902-1905. Promoted col. in 1903, retired from active service June 27, 1906 with rank of brig. gen. Appt. to Intl. Waterways Commission, 1903, Chair of American Section until 1913. Consulting engr. of the East Side Levee Assoc. and prepared plans for protection of the American bottom opposite St. Louis, MO. Consulting engr. on plans for drainage of Coldwater-Tallahatchie Basin in the alluvial bottom of the Mississippi River, 1909. Pres., bd. of consulting engrs. to examine plans to protect valley of Miami River, OH, from effects of floods, 1914-15. Given rank of maj. gen, Nov. 2, 1916. Dir. of Panama RR for over twenty years after retirement until a few months before his death. Publ: *Manual of Practical Military Engineering* (1873), *Report Respecting Tunnels under the Chicago River* (1904), *Report Upon Survey with Plans and Estimates of Cost for a Navigable Waterway from Lockport, Ill., to St. Louis* (1905), *The Preservation of Niagara Falls* (1906). Refs: DAB, TASCE, WWW, Lamb, CAB, WAB, BDNA.

ESTABROOK, JOHN DAVIS; b. Holden, MA, May 30, 1837; d. West Gloucester, MA, Jan. 26, 1918; f. John; m. Amanda; w. Frances Hildreth Mansfield; w. (2nd) Eliza Odiorne Mansfield; c. Mansfield and Francis Hildreth. Attended public schools and an academy. Entered RPI, March 1854, grad. in 1856 with the degree of civ. engr. Engaged in RR const. with the Louisville & Nashville RR Co. in Kentucky, 1856-59. Worked with the Council Bluffs & St. Joseph's RR, IA, 1859. Employed in offices of the city engr. of Boston, 1860-63. Supt. erection of coastal defenses in Cape Ann and Cape Cod, MA, 1863-64. During this time built Eastern Point Fort at Gloucester, MA. After the Civil War was engr. and mine supt. with the Little Schuylkill Navigation & Coal Co., Tamaqua, PA, 1864-68. Appt. engr. of public parks for Philadelphia, 1868-79. Dominant factor in the const. of Fairmount Park. Engr. and secy. of the Union Depot Co. in St. Paul, MN, 1879-82. In charge of const. work on the Northern Pacific RR, 1880. Chief commissioner of highways in the city of Philadelphia, PA 1883-84. Supt. of city parks in St. Paul, MN. Secy. and engr. in the flour mills of the C. C. Washburn Co., Minneapolis and held that position until he retired, 1899. Contributed to engr. publications and other magazines and newspapers. Publ: *Three Generations of Northboro Davises*. Mem: Civ. Engrs. Club of St. Paul (pres.), Philadelphia Engrs. Club, RPI Mining Engrs. Refs: RPI Bio, WAB.

EVANS, ANTHONY WALTON WHITE; b. New Brunswick, NJ, Oct. 31, 1817; d. NYC, Dec. 28, 1886; f. Thomas; m. Eliza (White); w. Anna Zimmerman; c. 2 sons and 1 daughter. Grad. RPI, 1836. Employed for seven years on the Erie Canal enlargement. Resident engr. on const. of the Harlem RR, 1845-50. Went to Chile to const. the Copoapo RR with Alan Campbell, 1850. Trip through Europe examining public works, 1853. Chief engr. in Peru of the Arica and Tacna RR. Declined position on the const. of the RR from Santiago to Valparaiso, 1856, and returned to the US. Went to Chile as chief engr. of the Southern RR, 1856-59. Chief and consulting engr. of the Copiapo RR extension. Returned to the US in 1860 for health reasons. Employed by the US govt. as engr. of the harbor defenses of NY, 1862-65. Commissioned by the Secy. of the Interior to establish standards for the RRs to the Pacific Coast, 1866. Purchasing agent for RR supplies for a large number of South American govts. and corporations and for the govt. of New Zealand, 1866-86. Consulting engr. in many important RR and building enterprises in those countries. Designed the Verrugas Viaduct on the Lima and Oroya RR in Peru, an iron bridge across a chasm 252 ft. deep and 500 ft. wide at an elevation of 5,836 ft. above sea level in 1872 and destroyed by floods in 1889. Mem: ASCE, Soc. of Civ. Engrs. of London, Am. Geographical Soc., Numismatic and Antiquarian Soc. of Philadelphia, and Soc. of the Cincinnati. Publ: *Supplement* to Edward Bates Dorsey's book, *English and American Railroads Compared* (1887), and articles in the *Transactions* of ASCE. Refs: WAB, DAB, CAB, RPI Bio.

FELTON, SAMUEL MORSE, SR.; b. Newbury, MA, July 17, 1809; d. Philadelphia, PA, Jan. 24, 1889; f. Cornelius; m. Anna (Morse); w. Eleanor Stetson; w. 2nd Maria Lippitt. Worked as clerk and bookkeeper to put himself through Livingston County High School, Genesco, NY. Entered Harvard 1830, grad. 1834 with degree of civ. engr. Taught in private school, 1836-38. Engr. for Loammie Baldwin, Jr. in Boston, 1836. Took over business after Baldwin's death, 1838. Built Fresh Water Pond RR to transport ice into Boston, 1841. Engr. of Fitchburg RR, 1843, supt. 1845. Pres. Philadelphia, Wilmington & Baltimore RR, 1851-64. Pres. of Delaware RR. Planned and directed secret passage of Abraham Lincoln from Harrisburg to Washington before his inauguration as Pres. in 1861. Commended by War Dept. for role in transporting Gen. Butler's troops to Annapolis, MD in April 1861, and preparing plans for cooperation of all RRs centering in Philadelphia. Commissioner of Hoosac Tunnel, 1862-64. Stroke causing paralysis forced retirement from active work, 1864. Returned to work as pres., Pennsylvania Steel Co., 1865-88. Commissioner of Union and Pacific RRs, 1869. Organizer, dir. of North Pacific RR, 1870-73, and of the Pennsylvania RR, 1873-83. Managing dir., Lehigh Navigation Co. Publ: "Philadelphia, Wilmington and Baltimore RR Investigation into the Alleged Misconduct of the Superintendent" (1854-55). Refs: DAB, CAB, WWW.

FELTON, SAMUEL MORSE, JR.; b. Philadelphia, PA, Feb. 3, 1853; d. March 11, 1930; f. Samuel Morse; m. Maria Low (Lippitt); w. Dorathea Hamilton; c. Hadassah, Ruth, Dorothy, and Samuel Morse. Educated in private schools in Philadelphia. Enrolled in engr. course at Pennsylvania Military Academy at Chester, PA. Worked during vacations as a rodman on the Chester Creek RR, 1868. Leveler and asst. engr. on the Lancaster RR. Entered MIT, grad. in 1873 with degree in civ. engr. Chief engr. of the Chester & Delaware

River RR const. by the Reading RR Co. Appt. gen. supt. of the Pittsburgh, Cincinnati and St. Louis RR Co., 1874-81. Gen. supt. of the Little Miami and Cincinnati and Muskingum Valley RR Co., 1881-82. Gen. mgr. of the NY and New England RR Co., 1882-84. Asst. to the pres. of the NY, Lake Erie, and Western RR Co. and gen. mgr. of the NY, Pennsylvania and Ohio RR Co., 1884. Vice pres. in charge of traffic, then vice pres. in charge of operation, 1885. Pres. of the East Tennessee, Virginia and Georgia RR Co., 1890. Pres. of the Alabama Great Southern RR and the Cincinnati, New Orleans and Texas Pacific RR co. When the co. went into receivership, appt. receiver of the property, 1893. Receiver of the Columbus, Sandusky and Hocking RR Co. and the Kentucky and Indiana Bridge Co. 1893-89. Pres. of the Chicago and Alton RR, 1899-1907. Engaged by the Commercial Club of St. Louis, MO, to report on a municipal bridge project. Elected pres. of the Mexican Central RR Co. and the Mexican American Steamship Co., Dec. 1907 until the govt. merged all the principal RRs into the National RRs of Mexico. Returned to the US in 1909 and became chair of the Tennessee Central RR Co. Elected pres. of the Chicago Great Western RR Co. in the fall of 1909. Chair, Bd. of Dir. in 1925. Co-receiver and pres. of the Pere Marquette RR Co., 1912-14. Consulting engr. and advisor to Chief of Engrs., US Army, June 1916. Organized troops for the War Dept. to operate RRs in Mexico in case the US would have to advance south of the border. Planned organization of a reserve force of RR men in the Corps of Engrs. Appt. Dir. Gen. of RRs, US Army, title later changed to Dir. Gen. of Military RRs. Trip inspection to France, 1918, traveling among all the battle lines. Vice pres. of Port and Harbor Facilities Commission of the US Shipping Bd., May 23, 1918, became acting chair and resigned as Dir. Gen. of Military RRs on Dec. 1, 1918. First civilian to receive Distinguished Service Medal of the US Army. Officer of the Legion of Honor awarded to him by the French govt. Appt. hon. advisor to the Army Industrial War College, mem. of War Dept Business Council, and Chair of the Committee on Military Affairs of the Assn. of Railway Executives. Operation of the Chicago Great Western RR Co., made a federal operation during World War I, returned to private operation, 1920. Took charge of the RR again, almost completed improvements when retired because of bad health. Commissioner of Lincoln Park, Chicago. Chair of the Engr. Commission of Cincinnati to select location for new water works. Dir. of Central Trust Co. of Illinois, Central Securities Co., People's Trust and Savings Bank of Chicago, Equitable Life Assurance Society, Chicago Surface Lines and Colonial Land Co. Hon. LL.D. from the Pennsylvania Military Academy and from Marietta College of Marietta, OH. Mem: ASCE, Executive Committee of the Assn. of Railway Executives, Western Assn. of Railway Executives, Corporation of MIT, Western Committee on Public Relations, Am. Railway Assn. (dir.), Western Railway Assoc. (pres.), Western Soc. of Engrs., Am. Railway Engr. Assn., Franklin Inst., Am. Legion, Civil Legion, Soc. of Am. Military Engrs., Veterans of Foreign Wars, Ohio Soc., Sons of the Revolution, Ohio Soc. of Colonial Wars, New England Soc. in NY, Commercial Club of Chicago, Chicago Club, Chicago Athletic Assn., Saddle and Cycle, Univ. Club of NY and Commercial Club of Cincinnati. Refs: TASCE, WAB.

FINLEY, JAMES; b. Ireland, 1756; d. Fayette County, PA, 1828. Judge of the Court of Common Pleas and Justice of the Peace in Fayette County. Member of the PA State Assembly for Fayette County, 1792. Invented bridge with a deck supported by iron chains suspended by the bridge's two towers. Completed bridge crossing Jacob's Creek on the road between Uniontown and Greensburgh, PA, 1801. Built number of bridges in New England and in the Delaware and Potomac Valleys, including bridge over the Potomac above Georgetown, completed 1807. During the next eight years, 40 bridges built using Finley's design, including Newburyport Bridge over channel of the Merrimack River, MA. Received a patent, June 17, 1808. Deacon of the Presbyterian Church. Refs: GEPT, CCI, Kemp.

FREEMAN, JOHN RIPLEY; b. West Bridgeton, ME, July 27, 1855; d. Providence, RI, Oct. 6, 1932; f. Nathaniel Dyer; m. Mary Elizabeth (Morse); w. Elizabeth Farwell Clark; c. Clarke, Hovey Thomas, Roger Morse, John R. Jr., Evert Wendell, Nathaniel Dyer and Mary Elizabeth Freeman. Early schooling in Lawrence, MA and Portland, ME. Entered MIT, 1872, grad. with degree in civ. engr., June 1876. Employed during vacations by Essex Co. in Lawrence, MA, employed after grad., 1876-1886. Principal asst. engr. and personal asst. to the co.'s chief engr., Hiram F. Mills, 1878-1886. Water Commissioner, Winchester, MA, 1882-86. Chief engr. and special inspector for Associated Factory Mutual Fire Insurance Co. of Boston, 1886-96. Chief of Inspection Dept. at Essex Co., 1890-96. Consulted in water power, municipal water supply and factory const. while in Boston. Engr. mem, MA Metropolitan Water Board, 1895-96. Moved to Providence, RI in 1896. Engaged by the comptroller of NYC to investigate new sources of water supply, 1899-1900. Submitted report presenting valuable experiments on flow over crests of model dam, a recomputation of yield of the Croton system and estimate of future water consumption of NYC. Civilian engr. mem., special bd. on gun carriage tests, War Dept., 1902. Mem. of Commission on Additional Water Supply of NYC, 1903. Investigated yield and quality of all available sources of supply, report became basis of subsequent work of Bd. of Water Supply. Chief engr. investigations, Charles River Dam, Boston Harbor, 1903. Consulting engr., Boston Metropolitan Park Commission on sanitary and drainage problems, 1903-1904. Mem. of RI Metropolitan Park Commission in 1904, and served as pres. of Providence Gas Co. Appt. consulting engr. to the Bd. of Water Supply of NYC, created to const. Catskill system, 1905. Planned water power developments at Feather River, CA, 1904-1905, and St. Lawrence River, Long Sault, 1905. On commission of three engrs. engaged to report on Los Angeles Aqueduct, to bring waters of Owens River 240 miles to Los Angeles, CA, 1906. Consulting engr. on Isthmian Canal Locks and Dams, 1906-7. Engaged in 1909 as consulting engr. on problem of new sources of water supply for Baltimore, MD. Consulting engr. to San Francisco, CA and planned the Hetch-Hetchy system, 1909-1912. Other cities which sought advice on water supply problems were: Nashua, NH; Denver, CO; Seattle, WA; San Diego, CA; and Mexico City, Mexico. Study of city planning for East Providence, RI, 1911. Dir. of RI Hospital Trust Co. and of National Bank of Commerce in Providence for ten years. Mem., Water Bd. to Chinese govt. on improvement of the Grand Canal and prevention of disastrous floods on the Yellow River and the Hwai River. Went to China, 1919. Active during World War I as mem. of Engr. Bd. of Review of the Sanitary District of Chicago and prepared program for regulation of the Great Lakes. Report included exhaustive studies of winter evaporation from the Great Lakes and of minor earth movements or tilting affecting the problem of lake levels. Mem. of visiting Committee, Bureau of Standards,

Washington, DC. Chair National Advisory Committee for Aeronautics. In 1923 gave $25,000 each to the ASCE, Am. Soc. of Mech. Engrs. and Boston Soc. of Civ. Engrs. for traveling scholarships in hydraulics for young engrs. and junior profs. Conceived idea of National Hydraulic Laboratory in Washington DC, active in furthering passage of necessary legislation in Washington and in advising on its completion. Hon. degree Sc.D. from Brown Univ., 1904; Tufts Univ., 1905, Sachs Tech. Hochschule (Germany) 1926, Univ. of Pennsylvania, 1927, Yale Univ., 1931. Mem: ASCE (hon. mem., Sept. 30, 1930, dir., 1896-1898, vice pres., 1902 and 1903, pres., 1922), Am. Soc. of Mech. Engrs. (pres., 1904-1905), Boston Soc. of Civ. Engrs. (pres., 1893), Providence Engr. Soc., National Academy of Sciences, Am. Academy of Arts and Sciences, Century and Engrs. Clubs of New York, Cosmos Club of Washington, DC and Univ. Club of Providence, RI. Publ: "New Experiments on Hydraulics of Fire Protection" (1889), "Experiments Relating to the Hydraulics of Fire Streams" (1890; Norman Medal, ASCE), "The Nozzle as an Accurate Water Meter" (1891; Norman Medal, ASCE), "Safeguarding of Life in Theatres" (1896), "Report Upon New York's Water Supply" (1900), "Report of the Commission On Additional Water Supply for the City of New York" (1903), "The Need of a National Hydraulic Laboratory for the Solution of River Problems," *Regulation of Elevation and Discharge of the Great Lakes* (1926), *Hydraulic Laboratory Practice* (1929), *Earthquake Damage and Earthquake Insurance* (1932). Refs: WAB, DAB, TASCE, WWW, Aldrich.

FRIZELL, JOSEPH PALMER; b. Barford, Quebec, Canada, March 13, 1832; d. Dorchester, MA, May 4, 1910; f. Oliver; m. Mary (Beach); w. Julia A. Bowes; c. Arthur B. and Alice M. Early education at Bennington, VT, and Richmond, Canada. Self-trained in engr. Worked in a cotton mill in Manchester, NH until 1854 and was asst. in city engr.'s office, 1854-56. Studied civ. engr. during his spare time. Asst. engr. in the office of the Locks and Canals Co. of Lowell, MA, 1857-61. Asst. civ. engr. during the Civil War engaged in const. fortifications on the Gulf coast, 1861-65. Returned to the Locks and Canals Co., 1866-67, assisted James Francis in the preparation of treatise on "Lowell Hydraulic Experiments." Practiced for a short time in Davenport, IA. Est. gen. consulting practice in Boston, 1870-78. Performed hydraulic investigations and reported on Minneapolis water-power in 1875, on damages from diversions on the Charles River, Waltham, MA and on the Neponset River and Mother Brook. Asst. civ. engr. to US Engr. Dept., 1878. Patented an air compressor operated by falling water. Made hydraulic investigations in the west, specifically on const. of reservoirs at headwaters of the Mississippi River and investigations at the Falls of St. Anthony. Mem. of the Water Commission in St. Paul, reported on water supply, 1881. Chief engr. of Bd. of Public Works, Austin, TX. 1890-92. Designed and began const. of the Austin Dam. Engaged in gen. consulting practice in Boston, 1893-1903. Retired, 1903. Mem: ASCE, Tariff Reform League. Publ: *Water Power, an Outline of the Development and Application of the Energy of Flowing Water* (1900), "Strains in Continuous Beams" (1872), "The Water Power of the Falls of St. Anthony," "The Old-Time Water Wheels of America," "The Storage and Pondage of Water" and other articles in scientific periodicals. Refs: DAB, TASCE, WAB.

GAGEL, EDWARD; b. Mt Hope, NY, Oct. 25, 1858; d. Feb. 11, 1931; f. Christian; m. Anna M. (Aulinger); w. Jennie Field Smith; c. Anna Smith and

David Edward. Educated in public schools and attended course in civ. engr. at Cooper Inst. in NYC. Draftsman and field engr. for Dennis and Mairs of NYC, engaged in gen. surveying, June 16, 1876-Dec. 1877. Employed by Brooklyn, Flatbush and Coney Island RR Co., Dec. 1877-March 1879. Draftsman and asst. engr., Metropolitan Elevated RR Co. of NYC, 1879-Jan. 1880. Draftsman of West Side & Yonkers RR Co. until March 1880. Leveler and draftsman on location survey for NY & New England RR Co., looking to extension to the Hudson River until Jan. 1882. Transitman for the Erie and Wyoming Valley RR Co., Jan-July 1882. Contractor's engr. on const. of the Pittsburgh, McKeesport & Youghiogheny RR, July-Sept. 1882. Draftsman with the NY Central RR Co., Sept.-Dec. 1882. Entered service of the NY, New Haven and Hartford RR Co., Dec. 6, 1882 until retirement. Worked up through positions of draftsman, 1882-85, asst. engr. 1885-91, district engr. in charge of the NY, the Maugatuck and the Highland Divisions, 1891-1900; asst. engr. in charge of the Western District, 1900-1904; principal asst. engr., 1904-1905. Chief engr., Oct. 1, 1905-Oct. 1, 1929. Appt. chief engr. of the Central New England RR Co., until Dec. 1926 when by merger with the New Haven Co. it lost its separate existence. Appt. chief engr. of NY Connecting RR Co., Sept. 3, 1918-Oct. 1, 1929. Retired Oct. 1, 1929. Chair of the Building Committee of the First Congregational Church of West Haven, CT also a Deacon and a trustee. On Bd. of Education of West Haven, pres. of its Park Bd., member of its paving commission and of its building improvement soc. Chair of its Fire Bd., mem. of New Haven Harbor Commission (pres., 6 years), dir. of the West Haven Bank and Trust Co. Mem: ASCE, Am. Railway Engr. Assn., NY RR Club, Connecticut Soc. of Cov. Engr. (hon. mem.), Engr. Club of NY, and Union League. Refs: WWW, TASCE, WWE.

GAILLARD, DAVID DU BOSE; b. Fulton, SC, Sept. 4, 1859; d. Baltimore, MD, Dec. 5, 1913; f. Samuel Issac; m. Susan Richardson (Du Bose); w. Katherine Ross Davis; c. David Saint Pierre. Lived with grandparents at Clarendon and attended private country school. Moved to Winnsboro to attend the Mount Zion Inst., 1872-74. Worked as clerk in a general store outside of school. Appt. to US Military Academy by competitive exam in 1880, grad. June 5, 1884, ranked 5th out of 31. Commissioned 2nd lt. of engrs., June 15, 1884. Student and instr. at Engr. School of Application, Willet's Point, NY, 1884-87. Engr. duty in FL, in charge of various surveys and harbor improvements at St. Augustine and Tampa and Withlacoochee River, 1887-91. Mem. of Intl. Boundary Commission for five years, est. line between Mexico and the US, 1891-94. At Ft. Monroe and on Washington aqueduct, 1895. In charge of water supply of Washington, DC, Dec. 1895-May 1898. Promoted capt. on Oct. 25, 1895. Survey of Portland Channel in Alaska for seven months. During Spanish-American War, chief engr. with staff of Gen. James F. Wade, April-June 1898. Appt. col. of 3rd Regiment of Volunteer engrs., mustered in at Jefferson Barracks, MO and was taken to Cuba, June 1898. Regiment returned to the US and mustered out at Ft. McPherson, GA, May 17, 1899. Asst. to engr. commissioner of District of Columbia, 1899-1901. Placed in charge of river and harbor improvements on Lake Superior with station at Duluth, MN, 1901-1903. Transfered to Dept. of the Columbia. Selected as mem. of initial Gen. Staff Corps. After short duty at Vancouver Barracks and St. Louis, promoted maj., April 23, 1904. Student in Army War College, 1904-1905. Gen. staff duty in Washington and in Cuba, Oct. 1906-Nov. 1907. In charge of dredging and excavation of the Panama Canal, 1907. Organized the Chagres Division, 23 miles of excavation from Gamboa to Gatun. In charge of the Central Division, 23 miles between the Atlantic and Pacific Locks including excavation through continental divide by the Culebra Cut. Promoted lt. col, April 1909. Collapsed on the job, July 26, 1913, died five months later. Name of Culebra Cut changed to Gaillard Cut, army post at Culebra named Camp Gaillard through executive order by Pres. Wilson. Publ: *Wave Action in Relation to Engineering Structures* (1904). Refs: WWW, DAB, WAB.

DAVID DUBOSE GAILLARD (1859-1913)

GIAVER, JOACHIM GOTSCHE; b. Gjovik, Norway, Aug. 15, 1856; d. May 29, 1925; f. Jens Holmboe; m. Hanna Birgit (Holmboe); w. Louise C. Schmedling; c. Astrid, Birgit, Erling, Finn J. and Einar. Tutored at home, grad. Trondhiem Technical College with the degree of civ. engr., 1881. Came to US in 1882, became draftsman for Northern Pacific RR Co. at St. Paul, MN, 1882-83. Draftsman for Shiffler Bridge Co., 1883-85. Chief engr., 1885-90. Designed bridges over Allegheny and Monongahela Rivers at Pittsburgh. Asst. chief engr. of the World's Columbian Exposition, 1891-93. In charge of perfecting the wind bracing in structural framework of the buildings there. Designed the three-hinge arch in the Liberal Arts Building, the largest in the world at that time. Engaged in gen. contracting, 1893-96. Became citizen of the US, 1896. Bridge designer for Sanitary District of Chicago, 1896-98. Designed various bridges over Chicago Drainage Canal. Chief engr. for firm of D. H. Burnham and Co., 1898-1915.

Engr. work in development of the skyscraper from the spread-footing foundation, cast iron column, wrought iron structural framework to the cassion foundation and structural steel framework. In charge of over 400 of the largest buildings in the US. Leader in securing bill for licensing of structural engrs. in Illinois, 1915. Decorated with order of Knight of St. Olaf, 1st class, by King Haakon of Norway. Consulting engr. in Chicago, firm of Giaver and Dinkelberg, 1916. Const. tallest building west of NY at the time, Jewelers Building in Chicago. Trustee of Norwegian American Hospital in Chicago. Mem: ASCE, Structural Engrs. Assn. (pres.), Western Soc. of Engrs. (dir.), Norwegian Engrs. Assn. (pres.), Engrs. Club, and mem. of several leisure clubs. Refs: TASCE, WWW, WAB, WWE.

GOTSHALL, WILLIAM CHARLES; b. St. Louis, MO, May 9, 1870; d. NYC Aug. 20, 1935; f. Daniel H.; m. Minnie (Wortmann); w. Adelaide von Rathgen. Early education from private tutors. Draftsman for the Metzger Iron Works of St. Louis. Later served as engr. in RR const. Worked for Missouri Electric Light & Power Co., St. Louis, made the first thousand-hour incandescent lamp test for duration and efficiency. Govt. overseeing work on banks of the Mississippi River, 1893. Electrified Union Depot RR Co. lines in St. Louis, one of first installations of the three wire system, 1894. Worked in Illinois, rebuilt the electric RR in Cairo, const. another in Belleville. Similar work in Marshalltown, IA, Muncie, IN, and for the Grand Ave. RR, St. Louis. Consulting engr. on const. work for the St. Charles RR in New Orleans, LA. Went to NY, supervised conversion of the Second Ave. RR from a horse-car into a conduit electric system, 1897-98. Pres. and chief engr. of the NY & Port Chester RR. Const. high speed electric line, the first built in the US on its own right of way and with no grade crossings, completed 1912. Connected with RRs in Europe, Asia, and Africa as well as the US. Major in the Corps of Engrs. during the World War I. Performed services in AK for the US National Park System. Organized and directed excavation in Palestine, 1925. Mem: ASCE, Am. Inst. of Electrical Engrs., Am. Geographical Soc., Adirondack League and several other leisure clubs and organizations. Publ: *Notes on Electric Railway Economics and Preliminary Engineering* (1903), and several contributions to periodicals. Refs: DAB, WAB, WWE.

GRAFF, FREDERIC; b. Philadelphia, PA, May 23, 1817; d. Philadelphia, PA, March 30, 1890; f. Frederick; m. Judith (Sawyer); w. Elizabeth Mathieu. Schooled in business, placed with a hardware firm. Began study of engr. with his father. Asst. engr. in the Water Dept. of Philadelphia, April 6, 1842-47. Chief engr. of the Dept., 1847-56, 1866-72. First chief engr. elected to the Water Dept. after it was made an independent bureau at the consolidation of Philadelphia, 1854. Directed reorganization of the Dept., combined district works with principal city works. Planned and directed const. of the Corinthian Ave. Reservoir on George's Hill, the reservoir in the city park and the rebuilding of the Fairmount Dam. Recommended establishing of a park on the Schuykill River, prepared maps and plans, 1851. East Side Park Lands purchased and dev. into the Fairmount Park System. Mgr. of Port Richmond Iron Works of Philadelphia, 1860-63. Assoc. with Henry R. Worthington of Philadelphia and NY, 1873-77. Consulting expert on water supply. Mem: ASCE (dir., pres., 1885), Engrs. Club of Philadephia (pres.), Franklin Inst. (vice pres.). Publ:

"Report of the Committee on the Preservation of Timber" (1885), *The History of the Steam Engine in America* (1876). Refs: DAB, WAB, WWW.

GRAHAM, CHARLES KINNAIRD; b. NYC, June 3, 1824; d. Lakewood, NJ, April 15, 1889; w. Mary Graham. Received liberal education. Entered Navy as a midshipman in 1841, served in the Gulf Squadron during war with Mexico, resigned in 1848. Studied law and engr. in NYC. Licensed to practice law in 1855. Employed as surveyor in laying out Central Park. Const. engr. at Brooklyn Navy Yard, built dry docks and landings, 1857. Appt. maj. of 1st Regiment of the Excelsior Brigade during the Civil War. Promoted lt. col. and then col. of the 5th Regiment of same brigade mustered in Oct. 15, 1861, designated as the 74th NY. Mustered out in April 1862, reappt. in May and served throughout the Peninsular campaign. Fought at Fair Oaks, Malvern Hill and other battles. Employed on recruiting duty until well enough to return to service in the field. Brig. gen. of Volunteers, March 1863. Commanded a brigade of the II Corps at Chancellorsville and Gettysburg. Captured at Gettysburg, sent as prisoner to Richmond, remained until exchanged in Sept., 1863. Joined Butler's Army of the James, assigned to command naval brigade and flotilla of army gunboats, Nov, 1863. Maj. gen. of Volunteers, March 13, 1865. Mustered out, Aug. 24, 1865. Practiced his profession in NYC. Chief engr. of the dock dept., 1873-75. Surveyor of the port, 1878-83. Naval officer of the port, 1883-85. Engr. for the NY Bd. of Commissioners for Gettysburg monuments. Refs: CAB, WAB, BDNA, DAB, WWW, Lamb.

GRANT, EDWARD MAXWELL; b. Dean's Corners, NY, Nov. 21, 1839; d. Belgrade, Serbia, Oct. 21, 1884; f. Joseph Fellows; m. Caroline A. (Maxwell); Educated at Poultney Academy, Poultney, VT. Entered RPI, 1857, grad. 1860. Went to TN as civ. engr. on Howe truss bridge building. Came north, 1861, recruited co. of calvary. 1st lt. of 1st Mounted Rifles. Went to Norfolk, VA, detailed as capt. of Corps of Engrs. Returned south as engr. of const. of RR bridges, 1865. Erected all bridges on the East Tennessee & Virginia RR and East Tennessee & Georgia RR extending from Lynchburgh, Va to Dalton, GA. Mem. of Mallory, Grant & Co., Macon GA, 1869. Designed and built wrought iron bridge on Montgomery & Selma RR over the Alabama River, 800 ft. long. Located and built narrow gauge RR in Alabama from Tuskgee to Chehaw, forming junction with Montgomery and West Point RR, 1871. Developed Wilson Patent for making merchant iron direct from ore, 1874. Erected furnaces at Wigan, England and Finland. Appt. brig gen. in Bulgarian Army. Acted as war correspondent for the *London Daily Times* from Belgrade, Serbia at Bucharest, Rumania in Russo-Turkish War. Appt. US vice consul gen. at Belgrade until his death. Refs: CAB, RPI Bio.

GREENE, CHARLES EZRA; b. Cambridge, MA, Feb. 12, 1842; d. Ann Arbor, MI, Oct. 16, 1903; f. James Diman; m. Sarah Adeline (Durell); w. Florence Emerson; c. Albert Emerson and Florence Wentworth. Attended Cambridge High School and Phillips Exeter Academy, entered Harvard and grad. A.B., 1862 and M.A., 1865. Assoc. with his older brother in manufacture of firearms, 1862-63. Volunteered for service in Union Army during last year of the Civil War. Clerk in Quartermaster Dept. of the Army, 1864. Commissioned as a 1st lt., regimental quartermaster in the 7th US Colored Troops stationed at Richmond VA, then at Indianola, TX, Jan. 1865-66. After the war, studied civ.

engr. at MIT, grad. B.S., 1868. Employed by Bangor & Piscataquis RR in ME for two years. Asst. engr. on US river and harbor improvements in Maine and NH, 1870-71. City engr. of Bangor, 1871-72. Prof. of civ. engr. at Univ. of Michigan, 1872-1903. Dean of school of engr. after 1895. Instrumental in additions of depts. of mining, 1874, and mech. engr., 1880. Chief engr. of the Toledo, Ann Arbor & Northern RR, 1879-81, designed and built trestle bridge over the Huron River at Ann Arbor, MI. Consulting engr. on bridge const. for the Wheeling & Erie RR, 1882 and const. of water works at Ann Arbor, 1885. Assoc. editor of *Engineering News*, 1876-77. Consultant at different times on the Washington Monument and Congressional Library Projects. Mem: ASCE, Michigan Assn. of Engr. and Surveyors (pres.). Publ: *Graphical Method for the Analysis of Bridge Trusses* (1876), *Bridge Trusses* (1879), *Arches* (1879), *Notes on Rankine's "Civil Engineering"* (1891), *The Action of Materials Under Stress or Structural Mechanics* (1897), and contributions to scientific journals. Refs: WAB, BDNA, Lamb, TASCE, WWW, DAB, CAB.

GREENE, DAVID MAXSON; b. Brunswick, NJ, July 8, 1832; d. Adams, NY, Nov. 9, 1905; f. Joseph Langford; m. Susanna (Maxson); w. Maria N. Skinner. Moved to Adams, NY, 1835. Early education in district schools and at Adams Seminary. Entered RPI, Oct. 1850. Completed 3-year course in one year, grad. 1851 with degree of civ. engr. Asst. to prof. of mechanics and physics at the Inst., 1851-52. Appt. to State Engr. Corps on enlarging the Erie Canal, spring 1852. Survey of the Cleveland, Lorain and Indiana RR, in OH, Sept. 1853-54. Appt. instr. in RR surveying at RPI and elected prof. of geodosy and topographical drawing, 1855-61. Sent to US Military Academy for special course in topographical engr., spring, 1856. Appt. 3rd asst. engr. in US Navy, spring 1861. Attached to the US frigate *Susquehanna*. Participated in attack and capture of forts at Port Royal and Hatteras Inlet. Engaged in blockade duty off the Atlantic Coast and in Gulf of Mexico. Detached from the *Susquehanna*, Sept. 1862, and ordered to US Naval Academy at Newport, RI, as asst. prof. of natural and experimental philosophy and instr. in steam engr. for three years. Ordered to duty as asst. to the chief of the Bureau of Steam Engr. in Navy Dept. at Washington, DC for 3 years. Appt. to Treasury Commission early in 1868 to examine and test devices for determining the product of distilleries, called the "Whiskey Commission". Chief asst. engr. of Albany in charge of US steamship *Narragansett* for a cruise in the West Indies, spring 1868. Cruise sent back to port for two weeks when yellow fever broke out in May. Detailed as 1st asst. engr. in charge of the *Frolic* in NY harbor in Sept, 1868. Reported for duty, rendering resignation from naval service, 1869. Returned to Troy, engaged in profession as civ., mech. and hydraulic engr. and as consulting engr. and expert. Employed to examine Ottawa River early in 1870 and report to the Canadian Parliament on probable effect of sawdust and mill refuse discharged by lumbermen at Ottawa. Examined plans for water supply at Troy, NY in 1871, advocated pumping system which was adopted. Appt. chief engr. of Walloomsac RR, consulting engr. of Ottawa City Water Works, Canada. Appt. chief engr. of Dansville Water Works, NY, 1873. School Commissioner of Troy from 1873-1876. Elected engr. of NY State Commission to test devices to substitute steam for animal power on the canals. Appt. engr. in charge of Eastern Division of the State Canals, Jan. 1874. In July of the same year, appt. Deputy State Engr. until 1877. Returned to Troy and resumed gen. engr. practice. Elected Dir. of the RPI, Aug. 1878-1891. Gen. engr. practice in eleven different states, DC and

Canada. Chief engr. of Troy Water Works extension, 1879 and 1880. Mem: ASCE, Am. Soc. of Naval Engrs., New England Soc. of Naval Engrs., Soc. of Naval Architects and Marine Engrs., Am. Assn. for the Advancement of Science, Intl. Assn. of Navigation Engrs., National Geographic Soc. (fellow), Soc. of Founders and Patriots of the US, Sons of the Am. Revolution, Military Order of the Loyal Legion of the US, Naval Order of the US, Post John A. Griswald, G.A.R, the Army and Navy Club of NYC, the Ionic Club of Troy and a Mason. Publ: *Notes on Steam Engineering* (1886), *Method of Laying out Easement Curves on Railroads* (1891), *The Fly Wheel as Regulator* (1890), *Manuscript Noter on Map Projections, Practical Hydraulics and Railroad Notes for the Use of Students*. Refs: WWW, TASCE, RPI Bio.

GREENE, GEORGE SEARS; Lexington, KY, Nov. 26, 1837; d. Dec. 23, 1922; f. George Sears; m. Martha Barrett (Dana); w. Susan Moody Dana; Entered Harvard College in 1856, left in 1857 without a degree. Studied engr. under his father. Asst. engr. on the Croton Aqueduct, on various RRs in Cuba and with several mining cos. on Lake Superior. Made topographical surveys of Westchester County and Long Island, NY, 1868-73. Introduced changes in surveying instruments. Engr. in chief of the dept. of docks, NYC, 1875-97. Designed and const. sea wall surrounding greater part of Manhattan Island and building of the wharves and piers in Chelsea Improvement on the North River between Charles Street and Twenty Third Street. Practiced in NY as a consulting engr., 1897-1922. Mem., bd. of advisory engrs. of the State Barge Canal, 1911-14. Mem: ASCE (dir., treas. and vice pres.), Am. Inst of Architects (hon. mem), Harvard Club and Century Assn. Refs, WWW, DAB, TASCE, BDNA, CAB, WAB, Lamb.

GREINER, JOHN EDWIN; b. Wilmington, DE, Feb. 24, 1859; d. Baltimore, MD, Nov. 15, 1942; f. John; m. Annie (Steck); w. Lily F. Burchell; c. Lillian Burchell, Gladys Houston. grad. B.S., Delaware College, 1880, civ. engr., 1884. Draftsman for Edgemoor Bridge Works in Wilmington, 1880-83, and Keystone Bridge Co. in Pittsburgh, PA until 1883-85. Draftsman for the Baltimore & Ohio RR Co., 1885-86, inspector of bridges, 1886-87, chief bridge draftsman, 1887-89, asst. bridge engr., 1889-91. Designing bridge engr. for the Philadelphia Bridge works, 1891-93. Returned to the Baltimore and Ohio RR Co., engr. of bridges, 1893-99, engr. of bridges and buildings, 1899-1905, asst. chief engr. 1905-1908. Examined buildings damaged in the Baltimore fire of 1904, helped to formulate new building code for the city. Engaged in private practice as professional consulting engr. with offices in Baltimore, 1908-retirement. Clients included Baltimore & Ohio RR, Erie and Norfolk & Southern RR, and state govts of PA and MD. Const. more than 300 bridges including, Kentucky & Indiana Terminal RR Bridge, Chesapeake & Ohio RR Bridge, Soldiers and Sailors memorial bridge at Harrisburg, PA, Baltimore and Ohio span over the Susquehanna. Chair of committee on iron and steel structures of Am. Railway Engr. Assn., 1907-1908. Secured adoption of first standard RR bridge specifications published in the US. Mem. of subcommittee on gen. engr. of Committee on Engr. and Education, Council of National Defense. Mem., American Railway Commission sent to Russia to inspect and report on Russian RR bridges during the World War I. Chair of port development commission of Baltimore, 1920-28. Mem. of the MD Waterfront Commission. Consulting engr. to the MD State Roads Commission, and as assoc. with the const. of the Pennsylvania Turnpike. Hon. Sc.D. from Univ. of Delaware in 1917 and D. Engr. from The Johns Hopkins Univ., 1937. Mem: ASCE (hon. mem.), Am. Railway Assn., Am. Soc. for Testing Materials, Am. Inst. of Consulting Engr., and leisure clubs. Publ: "What is the Life of an Iron Railroad Bridge" (1896; Norman Medal, ASCE), "Coal Piers on the Atlantic Seaboard" (1915; James Laurie Prize, ASCE), "Rolling Loads on Bridges," "The American Railroad Viaduct," "The Russian Railway Situation and Some Personal Observations." Refs: WAB, WWE, WWW.

GRINNELL, FREDERICK; b. New Bedford, MA, Aug. 14, 1836; d. Oct. 21, 1905; f, Lawrence; m. Rebecca Smith; w. Alice Brayton Almy; w. 2nd Mary Brayton Page; c. seven. Attended the Friends Academy, New Bedford. Entered RPI, 1852, grad. first in class with degree in civ. engr., 1855. Draftsman and mech. engr. for Jersey City, NJ locomotive works, fall 1855-summer 1858, and fall 1858-summer 1860. Asst. engr. on const. of Burlington & Missouri River RR, summer 1858. Began work as draughtsman at Corliss Steam Engine Co., Providence, RI, winter 1860. Elected as treas. of co. but actually employed as supt. mgr. of Jersey City Locomotive Works, built locomotives for Atlantic & Great Western RR Co., Jan. 1865-69. Appt. as supt. of motive power and machinery, fall 1865. Pres. of Providence Steam and Gas Pipe Co., Providence, RI, active business mgr. and mech. engr., 1869. Devoted great deal of time to the introduction of automatic sprinkler, secured about 40 patents for improvements on sprinklers and invented a dry pipe valve and automatic fire alarm system. Organized the Gen. Fire Extinguisher Co. in RI, OH and NC, 1893. Dir. of banks in New Bedford and Providence and textile manufacturers. Mem: Am. Soc. of Mech. Engrs. Refs: DAB, WWW, CAB, RPI Bio.

GROW, HENRY; b. Philadelphia, PA, Oct. 1, 1817; d. Salt Lake City, UT, Nov. 4, 1891; f. Henry; m. Mary (Riter); w. seven; c. thirty. Apprentice as a carpenter and joiner in PA. Supt. all bridges, culverts and other structures on Norristown and Germantown RRs. Became member of the Mormon Church, May 1842. Moved to Nauvoo, IL, May 15, 1843. Built barn for the patriarch of the Mormon Church and worked on the Nauvoo Temple. Fought in the Battle of Nauvoo, Sept. 19-21, 1846, escaped into IA. Traveled to NE, Oct. 1846. Moved with family to Missouri, fall 1847. Operated saw and grist mill and did carpentry work until 1851. Left for Utah Territory. Arrived in Salt Lake City on Oct. 1, 1851. Worked for one year on public works. Worked on the Old Mormon Tabernacle and the Social Hall, winter 1851. Built first suspension bridge in the territory across the Ogden River, 1853. Built the Sugar Works, 1854. Worked on the building of the two saw mills in Big Cottonwood Canyon, 1855. Built an additional mill at the head of the canyon near Silver Lake, 1857. Const. temporary buildings in Provo, UT in the event that the Mormons moved south, and built suspension bridge over the Provo River, 1858. Placed cotton and woolen machinery in grist mill at mouth of Canyon Creek. Built lattice truss bridges across the Weber and Jordan Rivers, 1861. Mill work, 1863-4. Began building great Mormon Tabernacle, 1865. Completed in 1867, first building designated by the ASCE as a National Historic Civil Engineering Landmark, 1971. Supt. const. of the Zions Cooperative Mercantile Inst. main building and warehouse, 1868. Appt. on a mission for Mormon Church to preside over PA, DE and MD, Nov. 1876-June 1877. Tore down the Old Tabernacle and built the Assembly Hall when he returned to Salt Lake City. Supt. buildings and scaffolding and hoisting apparatus for the carpentry for the Mormon Church including the Salt Lake Temple. Went east to examine improvements of paper mills, 1880. Returned and built mill, completed in 1883. Refs: Gessel.

HENRY GROW (1817-1891), MORMAN CARPENTER AND BUILDER

GRUNSKY, CARL EWALD; b. San Joaquin, CA, April 4, 1855; d. Jan. 9, 1934; f. Carl Albert Leopold; m. Clotilde Josephine Frederica (Camerer); w. Mattie Kate Powers; c. Carl Ewald, Kate Louise, Eugene Lucius, Clotilde. Attended Stockton, CA High School, grad. 1870. Grad. Realschule, Stuttgart, Germany, 1872-74, Polytechnikum, Stuttgart, 1874-77, grad. 1877. First work as a topographer with river surveying party of State Engr. Dept. of CA, 1878. Asst. and chief asst. state engr. of Calif, 1879-88. In private practice mostly in irrigation work, sewerage and drainage, 1888-99, located at Sacramento and San Francisco. River rectification and drainage problems as mem. Examining Commission on Rivers and Harbors of CA, 1889-90. Consulting engr. to Commissioner of Public Works of CA, 1893-94. Mem. of San Francisco Sewerage Commission, 1892-93. City engr., San Francisco, 1900-1904. Mem. of Isthmian Canal Commission, 1904-1905. Consulting engr., US Reclamation Service, 1905-1907. Private practice as consulting engr., NY and San Francisco, 1907-1909, San Francisco, 1910- . Engr.D., 1910, Polytechnicum, Stuttgart, Germany. D.Eng, 1924, RPI. Senior member C. E. Grunsky Co. Mem: ASCE, California Academy of Sciences (dir.), Commonwealth Club of California (pres.). Publ: *Valuation, Depreciation and the Rate-Base* (1916), *Topographic Stadia Surveying* (1917), *Public Utility Rate Fixing* (1918), *Ways to National Prosperity* (1929). Refs: WWW, WWE.

HARRIS, DANIEL LESTER; b. Providence, RI, Feb. 6, 1818; d. Springfield, MA, July 11, 1879; f. Allen; m. Hart (Lester); w. Harriet Octavia Corson; c. eleven. Family moved to Plainfield, CT when he was two. Attended district school there and the Plainfield Academy. Worked during vacations in his father's mill. Grad. Weslyan Univ. in 1838. Worked for Norwich & Worchester RR. Employed on a survey for the Erie RR, 1939-40. Asst. engr. with the Troy and Schenectady, 1840-43. Worked for contractor building the road between Springfield and Hartford. Mem. of Boody, Stone and Harris, 1845-79. One owner of the Howe Truss Patent. Designed bridges throughout the country. Const. 27 bridges for the Hartford, Providence & Fishkill RR including bridge over the Connecticut River. Dir. of the Connecticut River RR, 1855, pres., 1879. Examined and reported on the condition and safety of the bridges of the St. Petersburg & Moscow RR for the Russian govt., 1859. Govt. dir. of Union Pacific in 1869. Secy. and in charge of the Eastern RR Assn., 1866-78. Assisted in introducing the vacuum brake into England for the Assn., 1874. Mem. of the MA House of Representatives between 1857 and 1873. Mayor of Springfield, MA, in 1860. Refs: WWW, DAB.

HARROD, BENJAMIN MORGAN; b. New Orleans, LA, Feb. 19, 1837; d. Sept. 7, 1912; f. Charles; m. Mary (Morgan); w. Eugenia Uhlhorn; w. 2nd Harriet (Shattock) Uhlhorn. Prepared for college in local schools, New Orleans and Flushing, NY. Attended Harvard, grad. B.A., 1856. Studied engr. and architecture in New Orleans. Draftsman in US Engr. Office in charge of const. of forts and lighthouses on the Gulf Coast from Mississippi River to the Rio Grand; appt. asst. engr., 1858. Began practice in New Orleans as engr. and architect, 1859-61. Enlisted as private in Confederate Army, April 1861. Commissioned as lt. in artillery regiment, detailed on engr. duty. Served as brigade and division engr. in fortification and defense of New Orleans and Vicksburg, MS. Commissioned as captain of engr. troops in VA, engaged in defense of Petersburg and Richmond. Surrender at Appomattox, 1865. Resumed professional practice in New Orleans, 1865-77. Appt. Chief State Engr. of LA on board to const. system of levees to protect alluvial regions of the state from overflow, 1877-1880. Appt. engr. mem. of Mississippi River Commission, 1879-1904. Surveyed Mississippi River and tributaries; improved main streams from junction of the Ohio to head of the passes. Chief City Engr. of New Orleans, 1888-1892. Advisory engr. for drainage, sewerage and water works systems in New Orleans. In charge of drainage, design and const. 1897-1902. Appt. as one of the first US delegates to the Intl. Congress of Navigation, 1903, unable to attend. Appt. mem. of Panama Canal Commission by Pres. Theodore Roosevelt when type of canal was determined and charge of the work was transferred to the next commission, 1904-1907. Consulting engr. in New Orleans. LL.D. from Tulane Univ., 1906. Mem: ASCE (dir., 1892-1894, vice pres., 1895-1896, pres., 1897), Assn. of Harvard Engrs. (vice pres., 1909), Louisiana Engr. Society. Refs: WWW, DAB, TASCE, WAB.

HASWELL, CHARLES HAYNES; b. NYC, May 22, 1809; d. NYC, May 12, 1907; f. Charles; m. Dorothea (Haynes); w. Ann Elizabeth (Burns); c. 3 sons and 3 daughters. Early education at high schools in Jamaica, NY. Grad. Collegiate Inst. of Joseph Nelson. Entered engine works of James P. Allaire, 1825-36. Entered Navy as 1st chief engr. in 1836. Began design for engines and boilers of US steam frigate *Fulton*, Feb. 19, 1836. Designed and supervised building of machinery for naval vessels *Mississippi, Missouri,* and *Michigan,* 1839-42. Designed and operated 1st steam launch, 1837. Mem. of Navy Bd. that designed 6 steam frigates. Engr.-in-chief, US Navy, 1845. Drew up gen. order that defined duties and responsibilities of naval engrs., Feb. 26, 1845. First to introduce zinc in marine boilers and in bottom of iron vessels to arrest oxidation of boiler plates, 1846. Retired 1851 and began private practice, 1852. Served on Common Council in NYC 1854-58, last year as pres. Re-entered naval service during Civil War, chief engr. to expedition of Gen. Burnside to North Carolina, present at the bombardment and capture of Roanoke Island and at other naval events of the campaign. Engr. of NY State Quarantine Commission, designed and directed completion of Hoffman Island and its buildings in the lower bay of NY. Built merchant vessels, designed and const. ballasted crib, Hart Island. Was engr. Health Dept., Dept. of Charities and Correction. Trustee NY and Brooklyn Bridge, 1877-78. From 1898, const. engr. in charge extension and improvement Rikers Island, Long Island, Sound; asst. engr., Bd. of Estimate and Appropriation. Expert to examine and report on boiler plant in Chicago, summer 1904. Mem: ASCE (hon. mem.), Am. Soc. of Mech. Engrs. (hon. mem), Inst of Civ. Engrs. of Great Britain (hon. and life mem.), Boston Soc. of Civ. Engrs. (hon. mem.), Inst. of Naval Architects of Great Britain, Am. Soc. of Naval Engrs., Municipal Engrs. of NY, Am. Inst. of Architects, NY Academy of Sciences, NY Microscopical Soc., Soc. of Authors, the Engrs. Club of NY, Philadelphia Engrs. Club, Union Club of NY, NY Bd. of Councilmen 1855-58 (pres., 1858). Publ: *Mechanic's and Engineer's Pocket Book* (1842), *Mechanics Tables* (1854), *Mensuration and Practical Geometry* (1856), *Bookkeeping* (1860), "Reminiscences of Early Marine Steam Engine Construction and Steam Navigation in the United States of America from 1807-1850" (1898-1899); "Reminiscences of an Octogenerian" (1907). Refs: WWW, DAB, TASCE, Lamb, CAB.

HAUPT, LEWIS MUHLENBERG; b. Gettysburg, PA, March 21, 1844; d. March 10, 1937; f. Herman; m. Ann Cecilia (Keller); w. Isabella Christiana; c. Eva Ruth, Elsie Catherine, Walter Cromwell, Bessie May, Eleanor, Florence Belle, Susan Gertrude, Edna Schaeffer, Lewis Herman. Attended Philadelphia public schools and the Germantown Academy. Assisted father, Herman Haupt, in building Troy & Greenfield RR and the Hoosac Tunnel. Entered Univ. of Pennsylvania, 1861. Left at the end of freshman year to enter Lawrence Scientific School, Harvard. Appt. by Pres. Abraham Lincoln to US Military Academy, fall 1863, grad. 1867. 1st lt. of Engrs., June 17, 1867. Assigned to Corps of Engrs., conducted triangulation of Lake Superior until Feb. 1869 and then assigned to 5th Military District, TX. Resigned from the Army, Sept. 20, 1869. Asst. engr. and topographer in charge of surveys of Fairmount Park, Philadelphia, 1869-72. Asst. examiner in US Patent Office, Washington, DC, 1872. Resigned a few months later to accept chair of civ. engr., Univ. of Pennsylvania, 1872-92. In charge of hydrographic survey of Delaware River, 1873. Franklin Inst. Drawing School, 1874-79. Worked on triangulation of eastern PA for US Coast and Geodetic Survey, 1875-80. Secy. of Bureau of Awards, Centennial Exposition, 1876. Editor, *American Engineering Register,* 1885-86. Invented reaction breakwater for creating channels through ocean bars, 1887; Cresson Gold Medal from the Franklin Inst. in 1901 for this invention. Delegate and commissioner from PA to Nicaragua Canal Convention, New Orleans, 1892. Commissioner on the Lake Erie and Ohio

River Ship Canal, 1892. Engr. in charge of surveys for the Delaware-Raritan Canal, 1894. Appt. chair of the Columbia-Cauca Arbitration Commission, March, 1897. Awarded gold medal of the Magellanic Premium in 1897 by the Am. Philosophical Soc. for discoveries in physical hydrography and invention of system for harbor improvements. Mem. of the Nicaraguan Canal Commission, July 1897-99. Asst. judge, transportation features of Paris Exposition, 1900. Member of the Isthmian Canal Commission, June, 1899-1902. Hon. A.M. from Univ. of Pennsylvania, 1883, Sc.D., Muhlenberg, LL.D. Pennsylvania College. Mem: ASCE, Am. Assn. for the Advancement of Science, Am. Philosophical Soc., Am. Inst. for Mining and Mech. Engrs., Civ. Service Assn., Franklin Inst., Philadelphia Geographical Soc., Philadelphia Engrs. Club (pres.). Publ: *Engineering Specification and Contracts* (1878), *Working Drawings and How to Make and Use Them* (1881), *The Topographer--His Methods and Instruments* (1884), *The Physical Phenomena of Harbor Entrances* (1887), *Special Report on the Railway Plant of the Paris Exposition* (1889), *Canals and their Economic Relation and Transportation* (1890), *A Move for Better Roads* (1891), and several scientific articles. Refs: Lamb, WWW, WWE, BDNA, WAB, CAB.

HEBERT, PAUL OCTAVE; b. Bayou Goula, LA, Nov. 12 1818; d. New Orleans, LA, Aug. 29, 1880; f. Paul Gaston; m. Mary Eugenia (Hamilton); w. Cora Wills Vaughan; w. 2nd Penelope Lynch Andrews. Attended elementary schools, then Jefferson College, grad. in 1836 first in his class. Grad. also first in class at the US Military Academy, 1840. Commissioned 2nd lt. in the Corps of Engrs. Appt. asst. prof. of engr. at US Military Academy, 1841-July 21, 1842. Supt. const. of Mississippi River Defenses, 1843-45. Resigned from the Army, 1845. Chief Engr. of LA, 1845-47. Commissioned lt. col. in the 14th Infantry. Participated in all important battles of the Mexico City campaign. Retired after Mexican War. US Commissioner to the Paris Exposition, 1851. Mem. of the LA constitutional convention in 1852, elected governor of LA, 1853-56. Brig. gen. in the Confederate Army during the Civil War, in command in LA during 1861. Transferred to the Trans-Mississippi Dept., given command of Dept. of Texas, in charge of defenses of Galveston. Given command of sub-district of North LA, took part in battle of Milliken's Bend. After end of the war, resumed responsibilities in LA. Appt. State Engr., and commissioner and civ. engr. of the Mississippi levees, 1873. Refs: DAB, BDNA, DR, WAB, CAB, WWW, Lamb.

HENCK, JOHN BENJAMIN; b. Philadelphia, PA, Oct. 20, 1815; d. Montecito, CA, Jan. 3, 1903; f. George Daniel; m. Caroline (Speiss); w. Mary Ann Kirby; c. 2 sons and 1 daughter. Self- educated, entered Harvard, grad. first in class, 1840. Principal of Hopkins Classical School in Cambridge, MA, 1840-41. Prof. of Latin and Greek at the Univ. of Maryland, 1841-42. Similar position at the Germantown Academy, 1843-48. Offices of Felton & Parker, Civ. Engrs., 1848-50. Field work on Fitchburg RR. Left for partnership with William S. Whitwell, engr. in connection with const. of Cochituate Water Works. Opened offices in Boston for gen. engr. work, especially on street RRs there. Employed as engr. by the commission in charge of the Charles River Basin and Back Bay development. Continued business alone after Whitwell retired in 1859. In charge of filling in, laying out and paving the Back Bay District, 1856-61. Took charge of dept. of civ. engr. in newly est. MIT, 1865-81. After retirement, spent four years in Europe and upon his return, moved to Montecito, CA. Publ: *Field Book for Railroad Engineers* (1854). Refs: DAB, BDNA, DR, CAB, Lamb.

HENNY, DAVID CHRISTIAN; b. Arnhem, Netherlands, Nov. 15, 1860; d. Portland, OR, July 14, 1935; f. David; m. Berindina (Lorentz); w. Julia Antoinette Hermanie Wetzel; c. David, Frances Berindina, George Christian and Arnold Lorentz. Educated at the Polytechnic School of Delft, grad. with degree in civ. engr., 1881. Worked on RR, bridge and drainage work in Holland, 1881-84. Emigrated to the US, 1884. Engaged in gen. civ. engr. work; RR, water works and irrigation const. in IA, CO, UT, NY, AL, IL, SD, WI and OH, 1884-92. Gen. mgr. and chief engr. with the Exclesior Wooden Pipe Co., San Francisco, 1892-1902. Became US citizen, 1894. Gen. mgr. of the Redwood Manufacturing Co. of San Francisco, 1902-1905. Built first large factory in the US with individual motor drives at Pittsburgh, CA, 1902. Appt. supervising engr. for Pacific Division of the US Reclamation Service, Feb. 1905-1909 and then afterward as consulting engr. Work included dam const. in the West. Opened consulting office at Portland, OR, 1910. Mem. of several consulting bds for govt. and private agencies in the US, Cuba, Puerto Rico. for land reclamation, power, flood control and related problems. Invented Henny shear joint between the up stream and down stream faces of concrete blocks used in const. of massive concrete dams; designed while on Bd. of Consulting Engrs. for Boulder Dam on Colorado River, the highest built at that time. Patented invention and gave free use to the US Reclamation Service. Invention was employed in design and const. of Grand Coulee Dam on the upper Columbia River, containing greatest volume of concrete in any dam built until that time. Mem., consulting bd. concerned with proper location and design of Owyhee Dam in OR, Deadwood Dam in ID and Gibson Dam in MT. Consulting engr. to the Los Angeles County Flood Control District and chair of the Bonneville Dam Commission on the lower Columbia River. Mem. of the consulting bd. for Fort Peck Dam on the upper Missouri River. Chair of the ASCE special committee on irrigation hydraulics. D. Engr., Oregon State College, 1933. Mem: ASCE (dir., 1920-22, vice pres., 1932-33), Royal Inst. of Engrs. of Holland (hon. mem.), Am. Concrete Inst., Technical Soc. of the Pacific Coast (pres., 1900-1902), Oregon Soc. of Engrs. (pres., 1909-11), Oregon Tech Council (pres., 1920-22), Engr. Council, Advisory Committee of the Reconstruction Finance Corp. Publ: "Stability of Straight Concrete Gravity Dams" (1936; Norman Medal, ASCE). Refs: DAB, WWW, WWE, WAB.

HILDENBRAND, WILHELM; b. Karlsruhe, Germany, June 1, 1843; d. Feb. 21, 1908; w. Miss Hubbard. Classical education in the Lyceum of his town, entered the polytechnic school. Grad. in engr., passed "Staats Examen," entered govt. employment as engr. of highway const. and inspector of rails at the rolling mills of Westphalia and Rhenish, Prussia. Emigrated to the US in 1867. Draftsman in an architect's office. Employed with John Roebling in planning const. of the Brooklyn Bridge. Made the first drawings of the structure. Asst. engr. in the NY Central RR. Made architectural design for 42nd St. terminal station in NYC and const. of the great arched roof over the train shed, largest span in the world at the time. Principal asst. engr. in the actual const. of the Brooklyn Bridge begun in 1870. Made many scientific investigations and mathematical calculations necessary for the structure. Architectural design for the approaches and steel superstructure was designed, inspected and erected under his direction. Opened office in NY after completion of the bridge as consulting engr., 1883. Built several suspension bridges in the US and Mexico. Built truss bridge over the Ohio River at Wheeling, WV. Awarded second prize in submission to the public competition for the Washington Bridge, several of his features incorporated into the bridge. Appeared before a commission of five engrs. in NY to submit a design, estimate and argument in favor of a suspension bridge near the center of the North River, 1894. Chief engr. of reconst. of the Roebling Covington and Cincinnati Bridge over the Ohio River, 1895-99. Const. a light suspension bridge in Mexico for transport of silver ore from the mines of Penoles Co. at Mamimi to the terminal point of a rack RR. Const. track of Pike's Peak RR in Colorado as a representative of the ABT system of inclined rack RRs, 1889-90. Chief engr. for John A. Roebling's Sons Co., building cables of the Williamsburg Bridge, 1900. Consulting engr. of Westinghouse Electric Co. of Pittsburgh, 1904. Designed the overhead structures for electrification of passenger traffic of NY, New Haven and Hartford RR, near NYC. Also acted as consulting engr. of the co. until 1908. Mem: ASCE. Publ: *The Theory and Construction of Wire Cables, Underground Haulage of Coal by Wire Rope (*1888; awarded a medal and diploma of honor at the World's Fair in Chicago in 1893), book in German on the history of suspension bridges (1895). Refs: TASCE.

WILHELM HILDENBRAND (1843-1908)

HOADLEY, JOHN CHIPMAN; b. Martinsburgh, NY, Dec. 10, 1818; d. Oct. 21, 1886; f. Lester; m. Sarah (Chipman). Moved to Utica, NY in 1824, attended common schools and spent two years in a machine and pattern shop in Utica. Rodman on survey for RR between Utica and Binghamton. Returned to Utica Academy for one year on technical study. Engr. survey for enlargement of the Erie Canal, 1836. In charge of work between Utica and Rome, NY until 1844. Civ. engr. for Horatio N. and Erastus B. Bigelow, textile manufacturers.

Located, const. and installed new mills. Formed firm of McKay and Hoadley, Manufacturers and Engrs. in Pittsfield, MA, 1848. Const. mill machinery, steam engines and water wheels. Supt. and gen. agent of Lawrence Machine Shop in Lawrence, MA. Const. textile and paper mill machinery, water wheels, stationary steam engines, and locomotives, 1852-57. Representative in the MA legislature, 1858. Manufactured portable steam engines for twenty years. Four years of that time spent as dir. of the New Bedford Copper Co. Commissioned capt. in the MA militia, sent to England to inspect and report on ordnance for harbor defense for the state, 1862. In charge of const. with the McKay Sewing Machine Assoc., 1868. Consulting practice, 1873. Represented manufacturers and purchasers at tests of mill machinery and water works. Expert witness in patent and damage litigations. Organizer of the Clinton Wire Cloth Co. and pres. of the Archibald Wheel Co. Trustee of MIT. Mem: Am. Soc. of Mech. Engrs. (founder). Publ: several pamphlets incl., "The Portable Steam Engine" (1863), "Steam Engine Practice in the United States" (1884), also presented as paper to the British Assn. for the Advancement of Science; papers presented to the Am. Soc. of Mech. Engrs. included, "A Tilting Water Meter for Purposes of Experiment," "High Ratios of Expansion and Distriment," "High Ratios of Expansion and Distribution of Unequal pressures in Single and Compound Engines" and "Use of the Calorimeter as a Pyrometer for High Temperatures." Refs: DAB.

HODGES, HARRY FOOTE; b. Boston, MA, Feb. 25, 1860; d. Lake Forest, IL, Sept. 24, 1929; f. Edward Fuller; m. Anne Frances (Hammatt); w. Alma L'Hommedieu Reynolds; c. Antoinette, Frances, Alma Louise, Duncan. Educated at the Boston Latin School and the Adams Academy at Quincy, MA. Entered the US Military Academy, July 1, 1877. Grad. fourth in class, 1881. Commissioned 2nd lt. in US Army Corps of Engrs, June 11, 1881. Served for four years at the Engr. School and as Adjunct of the Battalion of Engrs. at Willit's Point, NY. Asst. to the district engr. in charge of works on the Great Lakes with headquarters at Detroit Michigan. Designed parts for the Poe Lick at the Sault St. Marie Canal and designed the steel lock gates, 1888. Asst. prof. of civ. and military engr. at the US Military Academy, 1888-92. Asst. to Col. Amos Stickney on the improvement of the Ohio and tributary. Promoted to capt., May 18, 1893. Placed in charge of improvements of the Upper Missouri and other rivers, 1893. Mem., Engr. Bd., originated design of features for mounting modern sea-coast guns. Appt. lt. col., June 10, 1898, and col., Jan. 21, 1899 of the 1st US Volunteers, Puerto Rican Division. Remained in Puerto Rico making roads and surveys and const. defensive works, reservoirs, refrigerating plants and bridges. Returned to the US, 1899 and mustered out, Jan. 25, 1899. In charge of river improvements at Cincinnati, 1899-1901. Chief engr. in the Dept. of Cuba in charge of works of the port of Havana and other military and civ. projects, 1901-1902. Assigned to office of the chief of Engrs. in Washington, DC, 1902-1907. Delegate to the 10th Intl. Navigation Congress at Milan, Italy. Promoted to the grade of maj. Gen. purchasing officer for the Isthmian Canal Commission, supervised the purchase of large quantities of materials for const. of the Panama Canal. Mem. of the Panama Canal Commission, Aug., 1907-1914. Asst. chief engr. of canal const., gen. supervision of the design and const. of the locks, dams and control works. Returned to the US after the first steamer passed through the canal, Aug. 15, 1914. Promoted to brig. gen. of the line of the Army, March, 1915. Received the thanks of Congress for his services, March

4, 1915. In charge of the Washington Engr. District until May 1915. Commanded Ft. Totten and the Middle Atlantic Coast Artillery District, 1915-17. Promoted maj. gen., National Army, Aug. 5, 1917. Command of the 76th Division, Camp Devens, MA, 1917. Supervised building of the cantonment, organization and training of the Division. Observation tour of the Allied Forces in France, Dec. 10, 1917-Feb. 1918. Resumed command of the 76th Division. Sailed to France, July 1918. Awarded the Distinguished Service Medal. Commanded the 20th Division at Camp Sevier, SC, Camp Travis, TX and the North Pacific Coast Artillery District. Promotted maj. gen. regular Army, Dec. 21, 1921. Retired on Dec. 22, 1921. Mem: ASCE. Publ: *Roster of Service with Engineering Troops of the United States Army, and a Brief Historical Sketch of Their Organization* (1885), *Notes on Mitering Lock Gates* (1892). Refs: TASCE, DAB, WWW.

HOGELAND, ALBERT HARRISON; b. Southampton, PA, Jan. 10, 1858; d. May 14, 1930; f. John; m. Keziah Delaney (Willard); w. Elizabeth T. Trego; c. Anna Trego. Attended public schools in Southampton and high school in Philadelphia, PA. Grad. Lafayette College with degree of civ. engr. in 1877. Rodman on surveying party for extension of the St. Paul & Pacific RR in MN and the Dakotas, April-Sept. 1879. Leveler on surveys for line of RR from St. Paul to Chicago, IL, Sept. 1879-March 1880. Leveler, transitman and topographer on location of main line of the Northern Pacific RR, Yellowstone and Rocky Mountain Divisions, April 1880-Feb. 1882. Asst. engr. for same company on the const. of the Bozeman Tunnel, Feb. 1882-March 1884. Asst. engr. on the St. Paul, Minneapolis & Manitoba RR on const. in ND, April-Sept, 1884. Asst. engr. for the Northern Pacific RR Co. on const. on the Wisconsin Division, Sept. 1884-June 1885. Employed with the St. Paul, Minneapolis & Manitoba RR, which soon became the Great Northern RR from July 1885-1930 in the following positions: asst. engr. on const. and maintenance on lines in MN, ND, SD and MT, July 1885-Sept., 1890; engr. maintenance of way in the Eastern District, Sept. 1890-Aug. 1896; resident engr. in the Eastern District, Aug. 1896-March 1902; asst. chief engr., March 1902-Feb. 1903; chief engr., Feb. 1903-1913; chair valuation committee, 1913-28; consulting engr., 1913-14; chief engr., 1914-25; consulting engr., May 1925-1930. Important works for the co. under his direction: const. of steel and concrete ore-handling docks and grain elevators at Superior, WI; electrification of the Cascade Tunnel and const. of a hydroelectric plant in WA; ocean docks and warehouses in Everett, WA and Vancouver, British Columbia; terminal improvements at Seattle including a double track tunnel under a section of the city; large freight and passenger terminals at Vancouver; a great number of steel bridges including a bascule bridge across Salmon Bay, Seattle and lift-span bridge across the Missouri and Yellowtone Rivers. Planned and supervised const. of an extension of the Great Northern RR Co. in MT later named for him. Chief engr. for the Glacier Park Hotel Co., a subsidiary of the Great Northern RR Co. Mem: ASCE, Am. Railway Engr. Assn., Engrs. Soc. of St. Paul, the Minnesota Club, the town and country clubs of St. Paul. Refs: TASCE, WWW, WWE, WAB.

HOLMES, HOWARD CARLETON; b. Nantucket, MA, June 10, 1854; d. Oct. 30, 1921; f. Cornelius; m. Maria (Folger); w. Josephine Bauer. Moved to San Francisco, 1860. Educated in public schools of San Francisco. Entered city engrs. office in Oakland. Topographic surveys and maps for the development of principal source of water supply for Oakland, Lake Chabot. Passed exam for appt. as US Deputy Surveyor. Asst. engr., Bd. of State Harbor Commissioners until 1883. Left position to design and build the Alameda Mole and Depot for the South Pacific Coast RR Co., 1883-92. Designed and built the Powell Street RR, 1887-88. Designed and built the cable RRs at Portland, OR, Spokane, WA and the Madison Street RR in Seattle, WA. In San Francisco, designed and const. the Sacramento Street Branch of the Ferries and Cliff House RR and lower end of the California Street RR. Extended the Union Street Cable RR from Fillmore Street to the Presidio. Designed and const. the original electric RR at Stockton, CA. Appt. chief engr. of the Bd. of State Harbor Commissioners of CA, 1892-1901. Built wharfs along water fronts at San Francisco and also the foundation of the Union Ferry Building at the foot of Market Street. Chief engr. of the San Franciso Dry Dock Co. to design and const. a 750-ft. granite and concrete graving dock at Hunter's Point, 1901. Consulting engr. for the Bethlehem Shipbuilding Co., designed and supervised const. of 1025-ft. concrete and granite graving dock also located at Hunter's Point, 1916. Chief engr. of San Francisco-Oakland Terminal RRs, designed and supervised the const. of ferry slips, terminal buildings, wharfs and extensive double-track trestle, 1902-1910. Consulting engr. during the const. of the Western Pacific RR, 1905-1910. Designed and supervised the Oakland Freight and Passenger Terminal and the San Francisco Freight Terminals of that co. Consulting engr. for the Panama-Pacific Intl. Exposition, 1913-15. Consulting engr. for the Esquimalt Shipbuilding and Drydock Co., Moore Shipbuilding Co., Richmond Belt RR, South San Francisco. Land and Improvement Co., Northwestern Pacific RR Co. and others. Mem: Engrs. Club, California Academy of Science, Masons, Knight Templar, Seismological Soc. of America, and leisure clubs. Refs: TASCE, WWW, WAB.

HOUGHTON, JAMES FRANKLIN; b. Cambridge, MA, Dec. 1, 1827; d. 1903; f. Charles; m. Mary (Briggs); w. Caroline Sparhawk; c. four. Attended high school in Waltham, MA. Entered RPI in 1846, grad. with degree of civ. engr. in 1848. Engaged in const. of Boston water works. Moved to CA in 1849. Mem. of shipping house of B. D. Baxter & Co., 1850-51. Dealer in lumber and private lands, 1851-84. Designed and built first state capitol at Benecia, 1852. Mem. of lumber firm of Pine & Houghton, 1853-4. Elected surveyor gen. of the state and ex officio register of state land office, 1862-68. Est. boundary lines through CA and NV. Pres. of Home Mutual Insurance Co., 1874-84. Presiding officer of Central Land Co. of Oakland, 1875-84. Pres. of the South San Francisco Dock Co., 1883-4. Pres. of the Corporation of Old Trinity Church and Parish of San Francisco. Mem. of the Bd. of Regents of the State Univ. of California. Dir. of the Pacific Mutual Life Insurance Co. Mem: Territorial Pioneers, California Academy of Sciences, Geographical Soc. of the Pacific, Academy of Sciences, San Francisco Art Assn., Astronomical Soc. of the Pacific and Pacific Union Club. Publ: *Report of the Surveyor General to the Governor and Legislature, Sacramento, California* (1862-68). Refs: WAB, RPI Bio.

HUMPHREYS, ANDREW ATKINSON; b. Philadelphia, PA, Nov. 2, 1810; d. Washington, DC, Dec. 27, 1883; f. Samuel; m. Letitia (Atkinson); w. Rebecca Hollingsworth; c. 2 sons and 2 daughters. Entered the US Military Academy, 1827, grad. July 1, 1831. Assigned to the 2nd Artillery, Ft. Moultrie, SC. On temporary duty at the academy, 1832. Served in GA and AL in the

Cherokee troubles, 1832-33. Served at Augusta Arsenal, GA and at Ft. Marion, FL, 1833-34. On topographical duty, West FL and Cape Cod, MA, 1834-35. In the FL war in the battles of Olokilikaha and Micinopy, 1836. Resigned from the Army, Sept. 30, 1836. Civ. engr. at Brandywine Shoal lighthouse and Cross Shoal breakwater, Delaware Bay, 1836-38. Reappt. to the Army on July 7, 1838 with rank of 1st lt. in Corps of Topographical Engrs. and served in the Topographical Bureau, Washington, DC, 1840-41. In the Florida War, 1842. At Washington, DC, 1842-44. In charge of the Coast Survey office, 1844-49. Commissioned capt., 1848. On survey in the field, 1849-50. On topographic and hydrographic survey of the delta of the Mississippi River, 1850-51. Examined protection of delta rivers from overflow in Europe, 1853-54. On duty in Washington, DC in connection with explorations and surveys for RRs to the Pacific Ocean and in geographical surveys west of the Mississippi River, 1854-61. Mem. of the Lighthouse Bd., 1856-62. Mem. of the Bd. at the US Military Academy to revise the program of instruction, 1860. Chief Topographical Engr. at Washington, DC, Dec. 1861-March 1862 and in the Army of the Potomac engaged in the defenses of Washington, the siege of Yorktown, the battles of Williamsburg, VA and movements and operations before Richmond, VA up to July 1862. Promoted maj., Corps of Topographical Engrs., Aug. 6, 1861. Col. of Volunteers, March 5, 1862. Appt. brig. gen. of Volunteers, April 28, 1862. Lt. col of Corps of Engrs., March 3. 1863. Maj. gen. of Volunteers, July 8, 1863 for action at Gettysburg, PA. Brig. gen. and Chief of Engrs., US Army, Aug. 8, 1866. Assigned to the command of the 3nd Division, 5th Army Corps at the battles of Antietam, Fredericksburg and Chancellorsville, VA. Chief of Staff to Gen. Meade, July 8, 1863-Nov. 25, 1864. Assumed command of the 2nd Corps, directed in the siege of Petersburg and the pursuit of Lee's Army to Appomattox. Commanded the District of PA from July 28-Dec. 9, 1865 to Aug. 8, 1866. Promoted maj. gen. by brevet in the regular Army, Aug. 8, 1866 for gallant and meritorious services at Sailor's Creek, VA. Commanded Corps of Engrs. and continued as Chief of Engrs., US Army until his retirement at his request, June 30, 1879. Served on Lighthouse and other bds. Mem: Am. Philosophical Soc., Hungarian Soc. of Engrs., Am. Academy of Arts and Sciences, National Academy of Sciences, Imperial Royal Geographical Inst. of Vienna (hon. mem), Italian Geographical Soc., Royal Inst. of Science and Art of Lombardy, Milan, Italy, Maryland Geographical Soc., Geographical Soc. of Paris, and Austrian Soc. of Engr. Architects. Hon. degree of LL.D. from Harvard, 1868. Publ: *Report on the Physics and Hydraulics of the Mississippi River* (1861), *The Virginia Campaigns of 1864-65* (1882), *From Gettysburg to the Rapidan* (1882), and contributions to biographical and scientific literature. Refs: BDNA, DR, WAB, CAB, WWW, DAB, Lamb.

HUNT, ROBERT WOOLSTON; b. Fallsington, PA, Dec. 9, 1838; d. Chicago, IL, July 11, 1923; f. Robert A.; m. Martha Lanchaster Woolston; w. Eleanor Clark; Educated in Covington, KY. Operated drugstore in KY, 1855-57. Worked in rolling mill of John Burnish and Co. at Pottsville, PA. Studied analytical chemistry, Philadelphia, 1859-60. Chemist, Cambria Iron Co., 1860-61. Est. first laboratory as a direct part of an iron or steel organization in US, Aug. 1, 1860. Capt. in command Camp Curtin, Harrisburg, PA, 1861-65. Recruited Lambert's Independent Mounted Co., 1864. Supt. steel works Cambria Iron Co., Wyandotte, MI, July 1865-May 1866. Moved back to Johnstown, PA for the co., 1866-73, completed first commercial order for steel rails for the Pennsylvania Steel Co. Assisted in design and erection of Cambria Bessemer steel plant and assumed charge on completion in July 10, 1871- Aug. 1873. Supt. of the Bessemer steel plant, John A. Griswold & Co., Troy, NY, Sept. 1, 1873-75. Gen. supt. of the Troy steel and Iron Co., March 1875-April 1888. Est. firm of Robert W. Hunt & Co. in Chicago, const. engrs., iron inspections, 1888-1923. Invented automatic rail mills. Trustee, RPI and hon. D. Engr., 1916. Secy. of committee which designed the ASCE rail section and the "A Section" of the Am. Railway Assn. Began special inspection involving thorough supervision of manufacture of steel and of the rolling of the rails. Proposed new rail section and the nick and break test for the soundness of each ingot, 1921. Awarded ASCE's John Fritz Medal, 1912, and the Washington Award in 1923. Robert W. Hunt Medal and the Robert W. Hunt Prize awarded by the Am. Inst. of Mining and Metallurgical Engrs. Mem: ASCE, Am. Inst. of Mining and Metallurgical Engrs. (hon. mem. and pres., 1883, 1906), Am. Soc. of Mech. Engrs. (hon. mem. and pres., 1891), Western Soc. of Engrs. (hon. mem. and pres., 1893), Canadian Soc. for Civ. Engr., Inst. of Civ. Engrs. of Great Britain, Inst. of Mech. Engrs., Iron and Steel Inst. of Great Britain, Am. Iron and Steel Inst., Am. Soc. for Testing Materials (hon. mem. and pres., 1912). Publ: "History of the Bessemer Manufacture in America" (1877), and "Evolution of the American Rolling Mill" (1892). Refs: WWW, DAB, TASCE, WAB, CAB.

INGERSOLL, COLIN MACRAE; b. New Haven, CT, Dec. 1, 1858; d. April 7, 1948; f. Colin Macrae; m. Julia (Pratt); w. Theresa McAllister; w. 2nd Marie Harrison; c. Theresa Manden Heuvel, Coline Macrae, Ralph McAllister. Primary education in Europe. Grad. Yale Univ., B.A., 1880. Asst. in the engr. dept. of the Missouri Pacific RR, 1880-81. Asst. engr. on the NY, New Haven & Hartford RR, 1881-1900, in charge of the double tracking of the Shore Line, improvements on the NY Division, and elevating the tracks through Boston. Asst. to pres. in charge at Boston, 1897-1900. Elected chief engr. 1900-1906. City engr., New Haven, 1892. Chair of the Harbor Commission of New Haven, 1895. Pres. of the Union Freight Co., Boston, 1897-1900, Old Colony Steamship Co., 1897-1900. Chief engr., Dept. of Bridges, NYC, 1906-1908, erected Manhattan and Queensboro Bridge and regulated traffic on the Brooklyn Bridge. Consulting practice, NYC, 1908-37, retired. Mem. of the Alaskan RR commission investigating transportation problems in Alaska, 1912-13. Mem: ASCE, Am. Inst. of Consulting Engrs., Delta Psi, Univ. Club of NY. Refs: BDNA, WWW, WAB, WWE, Lamb.

JACOBS, CHARLES MATTHIAS; b. Hull, Yorkshire, England, June 8, 1850; d. Laugharne, South Wales, Sept. 7, 1919; f. Bethel; m. Esther; w. Francis H. M. Fry. Tutored privately. Placed as a pupil in the firm of Charles and William Earle, Engrs. and Shipbuilders in Hull, 1866-71. Passed through workshops and drafting offices. Sent to China to put up bridges contracted by the co., 1871. Went to sea in the Merchant Service, obtained the British Bd. of Trade certificate as First-Class Marine Engr. Opened office as consulting engr. in South Wales, specializing in marine engr., 1876. Appt. surveyor to Lloyd's Register of Shipping at Cardiff. Moved to London in 1885, opened firm of Jacobs and Barringer, consulting engrs. Came to the US in 1889 to advise Austin Corbin, pres. of Philadelphia & Reading RR, on various projects, and designed tunnel for rapid transit between Brooklyn, NY and Jersey City. Opened office as consulting engr. in NYC, 1891-1916. Chief engr. of the East

River Gas Co., designed and const. tunnel, 10 ft. 8 in. diameter between Ravenswood, LI and 71st Street, NY, 1892-July 11, 1894. Invented the subterranean tunnel bridge, received a master patent from the US govt. and gold medal award at the St. Louis Exposition. Chief engr. of tunnel lines from Jersey City to NY, 1895-1910. Designed twin tunnels for NY and Jersey RR Co. from Hoboken to 42nd street and Lexington Ave., NYC. Designed and built two single track tubes to run from Jersey City Station of the Pennsylvania RR to a terminal at Church St. for the Hudson & Manhattan RR. Chief engr. of the Hudson Co., formed to const. tunnel lines of the NY & Jersey and Hudson and Manhattan RRs and to build extensions to make connection with the principal steam RR terminal on the NJ side of the North River, and with the suburban electric lines. Engaged in engr. works in England, India, and the US. Mem: ASCE, Inst. of Civ. Engrs. and Inst. of Mech. Engrs. of Great Britain, City of London and Royal Societies Clubs of London, Engrs. and Lawyers Clubs, NYC and Automobile Club of America. Refs: WAB, TASCE, WWW.

JOHNSON, EDWIN FERRY; b. Essex, VT, May 23, 1803; d. NYC, April 12, 1872; f. John; m. Rachel (Ferry); w. Charlotte Shaler; c. Louisa, Elizabeth, Edwin Augustus, William Shaler, Frederick Allen, Charles Shaler, Joseph Allen, and Lucy Ann Shaler. Family moved to Burlington, VT. Studied Latin with the Unitarian minister, taught land surveying by his father. Asst. in the Northeastern Boundary Commission, 1818. Survey of Lake Temiscouatta and route down that lake by the Madawasca and St. John Rivers to Madawasca settlement. In charge of astronomical observations and calculations. Entered Am. Literary, Scientific and Military Academy at Middletown, CT as student and tutor, Jan. 1823, grad. with honors, 1825. Instructor in mathematics and asst. prof. of natural history, 1825-26, prof. of mathematics and civ. engr., 1826-29. Accompanied the corps of cadets on their march to Plattsburg in 1824, Washington, DC, 1826, and Niagara Falls in 1827. Wrote a sketch of marches which were published in pamphlets. The school moved to VT, prof. of natural philosophy in the Weslyan Univ. in Middletown, 1829. Quit teaching. In charge of land surveys for the Erie Canal, 1829, Champlain Canal, 1830-31 and the Morris canal, 1831. Asst. engr. in charge of surveys for the Catskill & Canajoharie RR, 1831. Prepared plans and estimates for RR from Hartford to Guilford, CT, winter 1831-32. Supplied specifications and estimates to federal govt. for const. of a bridge over the Potomac River at Washington, 1832. Asst. engr. in charge of const. of the Chenango canal, April-Sept, 1833. Resident engr. on the Utica & Schenectady RR, Oct. 1833-Jan. 1835. Preliminary surveys for proposed Ontario and Hudson Ship canal during 1834. Principal engr. on Auburn Canal dam in 1835. Chief engr. of Auburn & Syracuse RR, 1835-38. Assoc. chief engr. on the NY & Erie RR, Feb. 1836-March 1837 and chief engr., March 1837-May 1838. Pres. of the Stevens Joint Stock Assoc. of Hoboken, NJ from July 1839-June 1840. Chief engr. of Ogdensburg & Champlain RR, 1840. Office as consulting engr. in NY in assoc. with W. R. Casey, 1843-45. Chief engr. of the NY & Albany RR, 1838-46. Chief engr. Syracuse & Oswego RR, July 1846-Jan. 1847. Chief engr. of the NY & Boston Air Line, 1847. Consulting engr. of Springfield & Boston Air Line RR, 1848-49, Rutland & Burlington RR, fall 1849. Made examinations of route and prepared plans and estimates for Vermont & Canada RR and the St. Lawrence and Champlain Canal, 1850. Plans and estimates for water works at Middletown CT, 1850. Consulting engr. on const. of a bridge at Wheeling, WV, July 1850. Chief engr. of the Rock River Valley Union RR, Sept, 1850-56, and

Illinois & Wisconsin RR, 1852-55. Reported on the const. of a RR from Troy to Oswego, March 1854. Survey of the city of Middletown and planned new system of sewerage, 1855. Compiled a new city charter, 1856. Mayor of Middletown, 1856-57. State senator, 1856. Pres. and treas. of Shaler & Hall Quarry Co., 1858-64. Conducted a cabinet and congressional party over the northeastern boundary, 1864. Examination for proposed ship canal and marine RR at Niagara Falls, July 1865. Consulting engr., Middletown water works, 1865. Surveys at Lewiston and Niagara Falls for ship canal, 1865-66. Consulting engr., Lake Ontario Shore Line RR, 1868-69. Appt. chief engr. of the Northern Pacific RR, June 14, 1866-70, then consulting engr. until his death, 1872. Invented improvement for canal locks, a screw power press, a six wheeled locomotive truck and an eight wheeled locomotive. Dir. of the Middlesex County Bank of Middletown and of the Shaler & Hall Quarry Co. Hon. degrees of A.M., Norwich Univ., 1829, A.M., Univ. of Vermont, 1839. Trustee of Norwich Univ., 1834-48. Publ: Treatise on Surveying (1825), Tyler's Arithmetic Revised and Reviewed (1827), The Newellian Sphere (1828), Land Surveys (1828), Review of the Project for a Great Western Railway (1829), Method of Conducting the Canal Surveys of New York (1832), Epicycloid (1832), Cubical Quantities, Railroads and Canals (1837), Mountains in New York (1839), Tables of Quantities for Tracing Railroad Curves (1840), Railway System of the State of New York (1840), Width of Track (1842), Gauge of Railways (1853), The Railroad to the Pacific, Northern Route, Its General Characteristics, Relative Merits, etc. (1854), Caesar's Bridge (1863), Report on the Defences of Maine (1862), Report of a General Plan of Operations to the Secretary of War (1863), Ship Canal and Marine Railways (1864), First Meridian (1864), Words for the People (1865), The Reciprocity Treaty (1866), The Navigation of the Lakes and Navigable Communications Therefrom to the Seaboard (1866), Niagara (1868), Water Supply of New York (1870), Transcontinental Railways (1870), Historical Sketch of Norse Settlements and the Newport Tower (1870), Banking and the Currency (1871), Broad and Narrow Gauge (1871). Refs: DAB, WWW, Norwich, WAB.

JOHNSON, JOHN BUTLER; b. Marlboro, OH, June 11, 1850; d. Pier Cove, MI, June 9, 1902; f. Jesse; m. Martha Butler; w. Phoebe Henry; c. 3 daughters and 2 sons. Prepared for college in public schools in OH. Family moved to Kokomo, IN, 1866. Attended one year of high school and Howard College. Attended normal school at Lebanon, grad. 1868. Taught for four years in elementary schools in IN and OH. Principal of the New London and Kokomo high schools. Went to Indianapolis as secy. of the School Bd. and instr. in the high school there, 1872-74. Entered Univ. of Michigan, 1874, grad. civ. engr., 1878. Civ. engr. on the US Great Lake Surveys for three years and asst. engr. for two years with Mississippi River Commission, 1878-83. Prof. of civ. engr. at Washington Univ., St. Louis, MO, 1883-99. Investigations into the strength of timber for the Division of Forestry of the US Department of Agriculture, 1892-95. Chosen dean of the College of Mechanics and Engr. at the Univ. of Wisconsin, 1899. Supt. Index Dept. of *Journal of the Association of Engineering Societies* from its organization in 1884, publ. 2 vols. of index notes to engr. literature, 1892. Mem: ASCE, St. Louis Railway Club, Western Soc. of Civ. Engrs., Intl. Assn. for Testing Materials, Am. Soc. of Mech. Engrs., Inst. of Civ. Engrs. of Great Britain, Engrs. Club of St. Louis, (secy., 1883), Am. Soc. for the Promotion of Engr. Education (secy. and pres.). Publ: *A Manual of the Theory* *and Practice of Topographical Surveying By Means of the Transit and Stadia* (1885), *Theory and Practice of Surveying* (1886), *Modern Framed Structures* (1892), *Engineering Contracts and Specifications* (1895), *Materials of Construction* (1897), and numerous articles in publications. Refs: BDNA, WAB, WWW, DAB, TASCE, Lamb.

JORGENSEN, LARS R.; b. Denmark, April 25, 1876; d. Berkeley, CA, May 8, 1938; f. Fritz; m. Patra Jansen; w. Karen Herskind; c. Ralph A. Served in Danish army six months before coming to US, 1901. Draftsman, General Electric Co., Schenectady, NY, 1901-1903. Draftsman and const. engr. on hydroelectric development, Edison Electric Co., Los Angeles, CA, 1903-1905. Hydroelectric engr., Pacific Gas and Electric Co., San Francisco, 1905-1907. F. G. Baumm & Co., 1907-14. Head of own firm, Constant Angle Arch Dam Co., San Francisco, 1914- . Invented constant angle arch dam, built first for a hydroelectric powerplant on the Salmon Creek near Juneau in southeastern AK. Mem. ASCE, AIEE, Engrs. Club of San Francisco. Refs: WWE, Schnitter.

JUDSON, WILLIAM PIERSON; b. Oswego, NY, May 20, 1849; d. Feb. 12, 1925; f. John Work; m. Emily (Pierson); w. Anna Littlejohn Thompson McWhorter. Educated in the public schools of Oswego, grad. from high school in 1866. Studied under his father and Capt. Jared Smith of the US Army Corps of Engrs. Civilian engr. with the US Corps of Engrs. on fort, river and harbor surveys with headquarters at US Engrs. Office at Oswego, 1866-98. Draftsman with Corps of Engrs. at Ft. Ontario, Oswego. Draftsman with fort and harbor const. on Lake Ontario, 1866-68. Surveyor and draftsman for forts, rivers and harbors, 1873-74. Appt. US engr. in executive charge of the surveys, design and const. of river and harbor works, lighthouses and forts on Lake Ontario and the St. Lawrence River including Lake Champlain, harbors of Buffalo, NY and Erie, PA, rebuilding and repair of Ft. Wayne, Detroit, MI, Fort Montgomery, Rouse's Point, NY, Fort Niagara, NY, and many surveys of projected works, including 10 miles of the Narrows of Lake Champlain. Reported on the Niagara Ship Canal, advising a barge canal from Oswego to the Hudson River instead of a ship canal, 1890. Survey and report for the US Deep Waterway Commission advising the canalization of the channels of these rivers, 1896. Asst. engr. to Maj. Thomas W. Symons, US Corps of Engrs. at Buffalo, 1897. Entered private practice and engaged as engr. for the contractor on const. of three miles of the Erie Canal at Buffalo, 1898. Deputy State Engr. of NY, 1899-1905. Executive charge of the organization and direction of the NY State Highway System during its first six years. Edited publications of the dept. Directed the surveying and monumenting of county lines through the Adirondack Mountains. Worked with the US Geological Survey in the survey of the state and the gauging of its streams. Consulting civ. and elec. engr., 1905-1925. Pres. of the Broadalbin, NY Electric Light and Power Co. and the Broadalbin Knitting Co., Ltd. and the Broadalbin Bank. Served on the Commission on the Varick Water Power Canal at Oswego since 1876. Mem: Inst of Civ. Engrs. of Great Britain, Am. Soc. of Municipal Improvements, MA Highway Assn., Intl. Navigation Congress, Am. Assn. for the Advancement of Science, Genealogical and Biographical Soc. of NY, NY State Hist. Assn., Am. Defense Soc., Sons of the Am. Revolution, Soc. of the War of 1812, and many leisure societies. Publ: *From the West and Northwest to the Sea By Way of the Niagara Ship Canal* (1890), *An Enlarged Waterway from the Great Lakes to the Atlantic* (1893), *Lake Ontario to the Hudson*

River Through the Oswego-Oneida-Mohawk Valley (1896), *History of Projects from Great Lakes to Tidewater* (1768-1901) *City Roads and Pavements* (1894; 1st ed.), *Road Preservation and Dust Prevention* (1908), and many engr. reports and papers. Refs: WAB, WWW, WWE, TASCE.

KIMBALL, GEORGE HENRY; b. Newburyport, MA, Dec. 8, 1849; d. ; f. Lafayette and Mary (Grover); w. Emma Carpenter; c. Ralph C., Wallace R., Harold L., Ruth, George H. Jr. Attended public schools in MA. Grad. 1873, MIT. Supt. of bridges and buildings, Pittsburgh, Cincinnati & St. Louis RR, 1876-79. Supt., Columbus & Sunday Creek Valley RR, 1879-80. Engr., maintenance of way, Little Miami RR, 1880-81. Chief engr., Toledo, Cincinnati & St. Louis RR, 1881-82. Supt., NY Chicago & St. Louis RR, 1882-89. Chief engr., LS & MS RR, 1889-91. In gen. practice as consulting engr., 1891-98. Supt. and chief engr., Columbus, Sandusky & Hocking Ry 1898-99. Chief engr., Pere Marquette RR, 1899-1902. Central Electric Const. Co., NY, 1902-1903. Engaged in designing a system of freight terminals for trunk lines at Buffalo, NY, 1903-1904. Chief engr., Chicago & Alton RR, 1904-1906. In gen. practice, 1906- . City commissioner, Pontiac Michigan, 1920-23. Mayor, 1923-24. Refs: WWE, WWW.

KINGMAN, LEWIS; b. North Bridgewater, MA, Feb. 26, 1845; d. in Mexico, Jan. 23, 1912; f. Issac; m. Sibel (Ames); w. Alice Newman; c. five. Educated in common schools and grad. Hunt's Academy, 1861. Entered three year course in civ. engr. with Shedd and Edson, Sept., 1862, for one and a half years. Engr. developing Wilkes Barre and Oil City, PA oil fields, 1863-68, with two months of engr. work in NYC. Began with Eastern Division of the Atlantic & Pacific RR, July 13, 1868. With Eastern and Western Divisions until Oct, 1871. Engaged on survey for the Maxwell Land Grant Co., 1871-72. On govt. land surveys in NM, 1873-76. Mountain surveys and const. in CO, NM and AZ for the Atchinson Topeka & Santa Fe, June 1877-July 1880. Worked on Atlantic & Pacific RR, July 1880-April 1883. Appt. chief engr. of the co. on Jan. 1, 1882. Chief engr. of Northern Division of the Mexican Central RR, April 1883-June 1, 1884. Chief engr. of the Atchinson, Topeka & Santa Fe RR, July 1884-Jan. 1, 1889. City engr. of Topeka, KS, 1889-94. Chief engr. of the Mexican Central RR, May 1894-1909. Engr. of maintenance of way, then office engr. Mem: ASCE, Am. Railway Engr. and Maintenance of Way Assn., National Geographic Soc., Am. Academy of Political and Social Science, the Franklin Inst. and Masonic Fraternity. Refs: TASCE, WWW.

KINGSLEY, WILLIAM; b. Fort Covington, NY, July 31, 1833; d. Feb. 21, 1885. Grew up on a farm. Taught school for a time. Supt. RR work in IL and WI. Contractor to const. city water-works in Brooklyn, NY, 1856. Shareholder in the NY Bridge Co. Supt. of work on bridge connecting Brooklyn with NYC. Placed in charge of a Bd. of Trustees, 1875. Succeeded Henry Murphy as pres. of the Bd. Refs: TASCE, CAB.

KITTREDGE, GEORGE WATSON; b. North Andover, MA, Dec. 11, 1856; d. Aug. 23, 1947; f. Joseph; m. Henrietta Frances (Watson); w. Georgia Davis; c. George Davis, Mary Henrietta. Educated in public schools of North Andover. Grad. MIT, B.S., 1877. With Metropolitan Water Supply, Boston, 1875. Private practice as civ. engr., 1878. Development and improvement of the

South Boston Flats Improvement, 1878-80. Maintenance of Way Dept. of the Pittsburgh, Cincinnati & St. Louis RR, connected with Pennsylvania lines west of Pittsburgh, PA 1880-90. Chief engr. of the Louisville Bridge Co., 1886-88. Engr. maintenance of way and asst. chief engr. Cleveland, Cincinnati Chicago & St. Louis RR, 1890-91. Chief engr., 1891-1906. Chief engr., Louisville & Jeffersonville Bridge Co., 1900-1906. Chief engr., NYC & H RR and Terminal Ry of Buffalo and of NJ Shore Line RR, 1906-14. Chief engr., NY Central RR, 1914-27. Chief engr., Hudson River Connecting RR, American Niagara RR, 1914-27. Chief engr., Beech Creek extension RR, 1911. Consulting engr., 1927-47. Mem: Corp. of MIT, 1907-12, Am. Assn. for the Advancement of Science, ASCE (vice pres., 1917-18), Am. Railway Engr. Assn. (pres.), and several leisure clubs. Refs: WWW, WAB, WWE.

KNEASS, STRICKLAND; b. Philadelphia, PA, July 29, 1821; d. Philadelphia, PA, Jan. 14, 1884; f. William; m. Mary (Honeyman); w. Margaretta Sybilla Bryan. Educated at the Classical Academy of J. P. Espy. Entered RPI, 1837, grad. 1839. Asst. engr. and topographer on the state survey for a RR between Harrisburg and Pittsburgh, PA. Draftsman with the Bureau of Engr., US Navy, 1840-42. Surveyed map of northeastern US-Canadian boundary, 1842. Asst. to chief engr. in const., Pennsylvania RR, 1847. Principal 1st asst. engr., 1848-53. Designed the first shop and engine house erected by the co. Assoc. engr., North Pennsylvania RR, 1853-55. Chief engr. and surveyor, City of Philadelphia, 1855-65. Designed new drainage for Philadelphia and bridges over the Schuylkill River. Made extended survey of the Susquehanna River from Duncan's Island to Havre de Grace, asst. in preparing topographical maps of the surroundings of Philadelphia. Asst. to pres., Pennsylvania RR Co., 1872-78. Pres., Eastern RR Assn., 1878. Pres., Belvidere Delaware RR Co., Trenton RR Co., Columbia & Port Deposit & Western RR Co., 1880-84. Dir., Pittsburgh Cincinnati & St. Louis RR, 1880-84. Mem: ASCE, Am. Philosophical Soc., Franklin Inst. Refs: RPI Bio, CAB, WWW.

KNOWLES, MORRIS; b. Lawrence, MA, Oct. 13, 1869; d. Nov. 8, 1932; f. Charles Edwin; m. Ellen B. (Richardson); w. Mina P. McDavitt; c. Helen Inez, Morris Jr. Grad. MIT, B.S., 1891. Apprenticed under Richard Hale, principal asst. engr., later chief engr. of the Essex Water Power Co. of Lawrence, MA. Surveyor in the Planning Department of the Assoc. Factory Mutual Insurance Co., 1890 and 1891. Asst. engr. for the East Jersey Water Co. Asst. engr. with MA State Bd. of Health, Sept. 1893- May 1897. Personal asst. in investigation of water supply for Boston, led to formation of the Wachusett Reservoir and Aqueduct to supply Boston and area with water. Mem. of the Water Bd. of Lawrence. Resident engr. of the Filtration Commission of Pittsburgh, PA, 1897-1910. Est. laboratory and experimental filters for improvement of the city water supply. Asst. engr. with John R. Freeman for preparation of report on additional water supply for NYC. Asst. engr. in charge of the Spring Garden and Torresdale Testing Stations in connection with improvement and filtration of water supply of Philadelphia, June 1900-July 1901. Chief engr. of the Pittsburgh Bureau of Filtration, Aug. 1901. Designed and supervised const. of the slow sand filters for the city's water supply on the Allegheny River, PA. Consulting engr. practice in partnership with L. E. Chapin in Pittsburgh, 1903-16. Designed impounding dam for development of an industrial water supply for the Tennessee Coal Iron and RR Co. in AL. Mem. of the engr. committee of the

Flood Commission of Pittsburgh, 1911. Mem. of the executive committee of the Flood Commission. Mem. of the Bd. of Advisory Engrs., Miami Conservancy District, 1914. Chief engr. of the Essex Border Utilities Commission, Canada, 1916-21. Supervising engr. in charge of const. of Camp Meade and Camp McClellan, 1917. Pres. and chief engr. of Morris Knowles, Inc. of Pittsburgh and Canada, 1916-32. Chief engr. of the Housing Division of the US Shipping Bd., 1918-19. Mem. of the Engr. Bd. of Review of the Chicago Sanitary District, 1924. Chair of its Committee on Const. Program. Appt. as mem. of the Advisory Committee in Zoning. Chair of the City Planning Commission of Pittsburgh, 1922-29. Chair of the Zoning Bd. of Appeals of Pittsburgh, 1923-27. Designed and supervised const. of sewage disposal and harbor improvements for Lynn, MA. Consulting engr. to the Atty. Gen. of Connecticut in investigations to restrain MA from the diversion of interstate waters in the extension of the Boston Metropolitan District water supply, 1928-30. Dir. of the Dept. of Sanitary Engr., Univ. of Pittsburgh, 1911-20. Honarary D. Engr. from the Univ., 1929. Vice chair of the Executive Committee of the Commission appt. by the Gov. of PA to study municipal consolidation in counties of the second class in PA. Drew up proposed charter for a federated city to amalgamate more than one hundred minor political subdivisions in Allegheny County into a Greater Pittsburgh. Chair of the Committee on Utilities in the Conference on Home Building and Home Ownership, 1931. Term member of the Corporation of MIT, 1924-29. Dir. of the National Conference on City Planning. Mem: Inst. of Civ. Engrs. of Great Britain, Verein Deutscher Ingenieure, Engr. Inst. of Canada, Am. Soc. of Mech. Engrs., Engrs. Soc. of Western PA (pres., 1923), Franklin Inst., Am. Water Works Assn., New England Water Works Assn., Pennsylvania Water Works Assn., Am. Public Health Assn., Am. Academy of Political and Social Science, National Inst. of Social Science, Boston Soc. of Civ. Engrs., National Municipal League, Duquesne Club, Univ. Club of Pittsburgh, City and Engrs. Clubs of NYC, Cosmos Club of Washington, DC, City Club of Boston, and Univ. Club of Philadelphia, Intl. Housing and Town Planning Assn., Sons of the Am. Revolution, Order of Founders and Patriots of America and leisure clubs. Publ: *Industrial Housing* (1920), and many articles in scientific journals. Refs: Morris Knowles, WAB, WWE, WWW, TASCE.

KOYL, CHARLES HERSCHEL; b. Amherstburg, Ontario, Canada, Aug. 14, 1855; d. Evanston, IL, Dec. 18, 1931; f. Rev. Ephraim Lillie; m. Frances (Culp); w. Georgiana Thatcher Washburn; w. 2nd Adele T. Sanford. Grad. A.B. from Victoria College, Coburg, 1877. Continued education at The Johns Hopkins Univ., 1879-81, fellow in physics after two years, 1881-83. Physical Science Master, Weslyan College, Quebec, 1877-79. Head of dept. of math and physics, high school in Washington, 1885-87. Delegate to Intl. Congress of Electricians, Philadelphia, 1884. Instr. in physics and electrical engr. at Swarthmore College, 1887-89. Patented parabolic semaphore for use in RR signaling, June 5, 1888. Awarded John Scoll Legacy Medal of the Franklin Inst. for this invention, 1889. Began engr. practice in NYC, 1890. Vice pres., then pres. of the National Switch & Signal Co., 1889-91. Pres. of the National Drying Co., 1891-93. Scientific asst. to the commissioner of street cleaning of NYC, 1895-96. Mgr., Automatic Banjo Co., NY, 1896-99. Mgr., Industrial Water Co., 1899-1902. Engr. of water service to lessen cost to the road of procuring nonalkaline water for use in the locomotive boilers, 1910. Engr. of water service for the Chicago, Milwaukee & St. Paul RR, 1920-31. Mem: St. Paul Soc. of Civ.

Engrs., Western Soc. of Engrs. Publ: "Municipal Refuse Disposal" (1908), "Prevention of Pitting in Locomotive Boilers by Exclusion of Dissolved Oxygen from Feedwater" (1929), "The Preparation of Water For Railroad Use" (1930), and other papers. Refs: WWE, DAB.

LANDRETH, OLIN HENRY; b. Addison, NY, July 21, 1852; d. Nov. 6, 1931; f. Rev. James; m. Adelia (Comstock); w. Eliza Taylor; c. William Comstock, Olin Henry Jr., Mary Eliza, Helen Adelia, James Taylor, Robert Nelson. Early education in high school at Rushville and at Penn Yan, NY, and at the Dundee and the Canesteo, NY academies. Grad. Union College with degree in civ. engr., 1876, and B.A., 1877. M.A., 1880. Sc.D., 1908. Various positions in RR engr., 1870-74. Assisted on various surveys, investigations and estimates of cost of water works at Schenectady, NY and at Olean, NY, 1874-77. Instr. of physics at Union College, 1876-77. Asst. astronomer of the Dudley Observatory at Albany, NY, 1877-79. Appt. prof. of civ. engr. at Vanderbuilt Univ., 1879-86. Supt. of Buildings and Grounds in charge of new const. and the maintenance and operation of the water supply, sewerage and steam heating systems. Surveys, estimates and investigations for RRs and water works and electric light systems in southern cities. Consulting engr. to the Bd. of Public Works of Nashville, TN, 1883-85. Consulting engr. to Davidson County Bridge Commission, to West Nashville Water Co. and to the TN State Bd. of Health, examinations on the sanitary effect of const. of proposed locks and dams on the Cumberland River. Dean of the engr. dept. at Vanderbilt Univ., 1886-89. Prof. of engr. at Union College, 1889-94. Special Commissioner to the World's Fair, Chicago IL, 1893. Mem. of the Advisory Council of the Engr. Congress at Chicago, 1893. NY State Water Storage Commission, 1902-1903. NY Bay Pollution Commission, 1903-1905. Metropolitan Sewerage Commission of NY, 1906-1908. Prof., civ. engr., Union College, 1894-1919. Leave of absence during World War I, served in an advisory capacity in the Ordnance Dept. as a mech. engr. in Washington, DC. Resigned from Union College at the end of the war. Mem. of the NY State Constitutional Convention of 1915. Consulting practice in NYC, 1919-31. Consultant to the Eastern Potash Manufacturers Assn. Review of extensive engr. undertaking in Poland for Am. financial interests. Olin H. Landreth Engineering Fund presented to Union College by the Class of 1926 for the advancement of engr. at the college. Mem: ASCE (dir.), Am. Soc. of Mech. Engrs. (vice pres.), Am. Inst of Consulting Engrs. (council, 1923-26), Am. Soc. for Testing Materials (pres.), Soc. for the Promotion of Engr. Education (dir.), Engr. Assn. of the South (secy.), Soc. of Engrs. of Eastern NY, Schenectady, NY Bd. of Trade (dir.), Special Joint Committee of Architectural and Engr. Societies on Structural Safety, Committee on Sanitation, Public Health and Water Supply of the Merchants Assn. of NY, Am. Assn. for the Advancement of Science, Sigma Xi Fraternity, Delta Upsilon Fraternity, Univ. Club of Mt. Vernon, NY. Publ: *Instructions for Finding the True Meridian with Tables for Surveyors, Metric Tables for Engineers* (1883). Refs: TASCE, WWW, BDNA, CAB, WAB, WWE.

LATROBE, CHARLES HAZELHURST; b. Baltimore, MD, Dec. 25, 1833; d. Baltimore, MD, Sept. 19, 1902; f. Benjamin Henry; m. Maria Eleanor (Hazelhurst); w. Letitia Breckenridge Gamble; w. 2nd Rosa Wirt Robinson; w. 3rd Louise McKim; c. 2 daughters and 1 son. Attended St. Mary's College, Baltimore, MD, studied also with his father. civ. engr. for Baltimore & Ohio RR, 1850. Chief engr. in charge of const., Pensacola & Georgia RR, Fla. Lt. of engrs., Confederate Army, 1861, completed grading, bridge building and rail laying on the last twenty miles of the Pensacola & Georgia RR. Assoc. with his father and Charles Shaler Smith in Baltimore Bridge Co., 1866-77. Engr., Jones Falls Commission, Baltimore, 1875-89. Designed and const. retaining walls, several iron bridges, in charge of the improvements and extensions in Mount Royal, Druid Hill and Patterson Parks, laid out terraced gardens along Mount Royal Ave. Commissioned by Peruvian government to construct bridge at Verrugas on Callao-Oroya-Huancayo RR, highest bridge in the world at that time, and the Arequipa Viaduct. Consulting engr. for several RRs. Refs: WWW, Lamb, DAB, WAB, BDNA, CAB.

LAURIE, JAMES; b. Bells Quarry, Scotland, May 9, 1811; d. Hartford, CT, March 16, 1875; never married. Apprentice in office of mathematical and engr. instruments in Bells Quarry until 1832. Worked for a year in office of a civ. engr. Came to MA as assoc. engr. with James P. Kirkwood who was chief engr. of const. of the Norwich & Worcester RR, 1833. Chief engr. and supt. of const., 1835. Consulting engr. for RR, canal, dam bridges and wharf companies. One of a group of engrs. who founded the Boston Soc. of Civ. Engrs., July 1848. Chief engr. of New Jersey Central RR. Plans for extension of the road from Whitehouse to Easton, 1849. Moved office to NYC, 1852-58. Founder, ASCE, Nov. 5, 1852, elected as first pres. Presented paper at the first meeting "The Relief of Broadway" calling for elevated RR tracks. Meetings not held from 1855-1867. Relinquished title on Oct. 12, 1867 when ASCE was revived. Examined RR bridges for the state of NY, 1855-56. Engaged by the govt. of Nova Scotia to examine and report on the condition of the Nova Scotia RR, 1858. Chief engr., 1859-60. Reported on Troy & Greenfield RR for the state of MA, 1862, and employed for several years as consultant on the Hoosac Tunnel. Chief engr. of the New Haven, Hartford & Springfield RR, 1861-66, designed and built its bridge over the Connecticut River at Warehouse Point. Examination of the Lyman Viaduct on the Air Line RR, 1870. Examined the Eads Bridge at St. Louis for the bondholders. Refs: DAB, TASCE, CAB, WAB, WWW.

JAMES LAURIE (1811-1875), FOUNDER AND FIRST PRESIDENT OF ASCE, 1852-1867

LEWIS, WILLIAM GASTON; b. Rocky Mount, NC, Sept. 3, 1835; d. Goldsboro, NC, Jan. 7, 1901; f. John Wesley; m. Catherine Ann (Battle); w. Martha E. Pender; c. James Spencer, Kemp Battle, William Gaston Lewis, Jr, Anna Hartwell, Laura Lloyd, Mittie Pender and Elizabeth Mason Lewis. Attended Lovejoy's Military School at Raleigh. Grad. Univ. of North Carolina, A.B. 1855. Taught at Chapel Hill, NC, and Jackson County, FL. Govt. surveyor in MN, 1857-58. Asst. engr. on Tarboro branch of Wilmington & Weldon RR, 1858-61. Mem. of the Edgecombe Guards forming part of the 1st NC Regiment, was made ensign and lt., April 21, 1861. Promoted maj. of the 33rd NC Regiment for service in the battle of Big Bethel. Promoted lt. col. of the 43rd NC Regiment, Jan. 17, 1862, for the battle of New Berne. Participated in Ewell's Shenandoah Valley campaign, June, 1863. Engaged in the battles of Malvern Hill and Gettysburg. Took part in the battle of Bristow Station, Mine Run and in the capture of Plymouth, NC, April 1864. Succeeded to col. Commended for const. of outer works at Drewry's Bluff, May 16, 1864. Promoted brig. gen. June 2, 1864 as of May 31, 1864, and assigned to Hokes, NC brigade. With Gen. Early in the campaign in the valley of VA and at Petersburg. Severely wounded at Farmville, April 7, 1865 and taken prisoner. After the war, roadmaster of the Wilmington & Weldon RR, 1865. Asst. engr. with the W. C. & R. RR, 1866. Supt. of the Raleigh & Gaston RR, 1867. Chief engr. of the Winston & Tarboro and the Edenton & Norfolk RRs, 1868-70. Merchant and farmer until 1885. Chief engr. of the NC National Guard, 1885-91. Agent of the NC swamp and phosphate lands. Chief engr. of the Albany & Raleigh RR, 1890. Refs: Lamb, DAB, WAB, BDNA.

LOVELL, MANSFIELD; b. Washington, DC, Oct. 20, 1822; d. NYC June 1, 1884; f. Dr. Joseph; m. Margaret (Mansfield); w. Emily Plympton. Educated in ordinary schools, entered US Military Academy, grad. 1842. Commissioned 2nd lt., 4th Artillery US Army, 1842. Served in the occupation of TX, 1845-46. Commissioned 1st lt., Feb. 16, 1847, asst. adjunct gen. of division of Gen. John A. Quitman. In war with Mexico, brevetted capt. for gallantry at Chapultepec, Sept. 13, 1847. Severely wounded at the battle of Monterey, Sept. 14, 1847. On garrison duty in several locations, 1849-54. Resigned from the Army, Dec. 18, 1854, to take part in Gen. Quitman's plan for Cuban Expedition. Expedition failed. Employed in Cooper & Hewitt's Iron Works, Trenton, NJ, 1854-58. Supt. of street improvements, NY, April, 1858. Deputy street commissioner in Nov. 1858-Sept. 1861, when he joined the Confederate Army. Appt. maj. gen. in the Confederate Army and ordered to command at New Orleans, Oct. 7, 1861. Withdrew forces on April 23, 1862. Commanded the I Corps in the Battle of Corinth, MS, Oct. 3-4, 1862. Relieved of command, Dec. 1862. Absolved of blame for the loss of New Orleans, Nov. 1863. Volunteer staff officer under Joseph E. Johnston, summer 1864. After the war, tried rice planting. Asst. engr. under Gen. John Newton in removing East River obstructions at Hell Gate, NY. Refs: DAB, WWW, BDNA, DR, CAB, WAB.

LUDLOW, WILLIAM; b. Islip, NY, Nov. 27, 1843; d. Convent Station, NJ, Aug. 30, 1901; f. William Handy; m. Frances Louise (Nicoll); w. Genevieve Almira Sprigg; c. 1 daughter. Educated at home and at Burlington, NJ. Univ. of the City of New York, 1853-60. Entered US Military Academy, July 1, 1860, grad. June 13, 1864, promoted 1st lt., Corps of Engrs. Chief engr., 20th Army Corps, July 19-Sept. 1864. Promoted capt. for the battle of Peach Tree Creek,

July 20, 1864. Participated in siege and capture of Atlanta, July 22-Sept. 2, 1864. In charge of const. of defenses at Rome, GA, Oct. 4-Nov. 15, 1864. In the engagement of Allatoona Heights, Oct. 5, 1864. Chief engr. of the Army in GA, Nov. 15, 1864-March 20 1865. Brevetted maj., Dec. 21, 1864. Asst. chief engr. to Maj. Gen. W. T. Sherman's march to the sea which ended with the surrender of Savannah, GA, Dec. 21, 1864, the invasion of the Carolinas, Jan.-March 1865. Engaged in the battles of Averysborough, March 16 and Bentonville, March 19, 1865. Promoted lt. col. for services in the Carolinas, March 13, 1865. At the occupation of Goldsboro, NC, March 22, 1865 and at capture of Raleigh, NC, April 13, 1865. Leave of absence, April 25-Nov. 16, 1865. Organized the engr. depot at Jefferson Barracks, MO, Nov. 19 1865-Sept. 4, 1866. Commanded engrs. at Jefferson barracks and in charge of engr. property in MO and AR, Dec. 12, 1865-Nov. 1867. Capt. of engrs., March 7, 1867. Asst. to Gen. Gillmore in charge of fortifications and river and harbor work at NYC and along the South Atlantic coast, 1867-Nov. 10, 1872. Chief engr. in the Dept. of Dakota, Nov. 10, 1872-May 9, 1876. Surveys of the Yellowstone National Park, 1873 and 1875, and of the Black Hills country, 1874. Asst. engr. to Lt. Col. Kurtz, 1876-77. Asst. engr. under Col. Macomb, 1877-81. Gen. engr. service on Delaware Bay and River, harbor and river improvements, fortifications and other work, 1877-82. River and harbor work in Philadelphia, 1881-82. Maj. of engrs., June 30, 1882. Engr. secy. of the Lighthouse Bd. at Washington, DC, Aug. 28, 1882-March 8, 1883. Chief engr. of the Philadelphia Water Dept., April 1, 1883-April 1, 1886, reorganized and improved the city's water system. Appt. engr. commissioner of the District of Columbia, April 1, 1886-Jan. 27, 1888. Engr. of the 4th Lighthouse District, March 1-Dec. 14, 1888. In charge of river and lighthouse work on the Great Lakes, 1888-93. Military attache US Embassy, London, 1893-96. Inspected deep water canals of Suez, Corinth, Kiel and the Netherlands. Chair of the Nicaragua Canal Bd., April-Nov. 1895. Lt. col., Corps of Engrs., Aug. 13, 1895-Feb. 23, 1897. In charge of river and harbor improvements in NY Harbor, 1897-98. Recommended that the East River channel be deepened. In war with Spain, appt. brig. gen. of Volunteers, May 4, 1898. With Shafter's V Corps in Santiago-de-Cuba, June 28, 1898. Commanded the 1st Brigade in the attack on El Caney and in the subsequent investment of the city of Santiago. Maj. gen. of Volunteers, Sept. 7, 1898. Pres. of bd. to organize Army Sea Transport Service, Sept.-Oct., 1898. Military governor of Havana, Dec. 13, 1898-May 1, 1900. Commissioned brig. gen. in the regular army, Jan. 21, 1900. Pres. of the Army War College Bd., 1900. Inspected French and German military systems and methods of training, summer, 1900. Active duty in the Philippines, April-May 1901. Sick leave absence May-Aug. 1901. Mem: ASCE, Companion of the Military Order of the Loyal Legion of the US. Publ: *Explorations of the Black Hills and Yellowstone Country.* Also *Report of the US Nicaraguan Canal Commission.* Refs: DAB, WWW, Lamb, BDNA, WAB, TASCE.

LUNDIE, JOHN; b. Arbroath, Scotland, Dec. 14, 1857; d. NYC, Feb. 9, 1861; f. James; m. Anne (Honeyman); w. Iona Oakley Gorham; w. 2nd Alice Eddy Snowden; c. none. Graduated from Dundee high school, 1873. Served in office of harbor engr. of the Port of Dundee, 1873-77. Entered Univ. of Edinburgh, finished 4 year course in 3 years, grad. 1880 with B.S. in engr., 1st prize in mathematical physics. Came to the US after grad., began RR work in Oregon and Washington Territory, 1880-84. In charge of building of the Table Rock Tunnel. Engaged in private practice in Chicago, IL, 1884-85. Became citizen of the US. Asst. engr. in charge of work in Chicago, 1886-90. Made preliminary survey for Chicago Drainage Canal, designed several of the bridges in the city. Principal asst. engr. in last year and a half. Engr. and gen. agent in Chicago for the King Bridge Co. of Cleveland, 1890-93. Designed and erected structural steelwork of four of World's Fair buildings and Illinois Central Train Shed at Chicago. Engaged in private practice in Chicago, 1893- . Designed 1st low level drainage system in city, and sewerage system of Kenosha, WI. Electrical engr. work for the Chicago and Worth Electric RR Co. Investigated and reported on an artesian water supply for Memphis, TN, 1895, developed method for determination of the yield of artesian water areas. Visited Isthmus of Panama to make survey and examination of Pacific Mail Co. property at Colon, 1896. Investigated and reported on application of electricity to suburban train service of the Illinois Central RR, 1896-97. Prepared a thesis from investigation focused on rapid acceleration, est. Lundie formula for train resistance and power rating of electric RR motors, received degree of D.Sc. from Univ. of Edinburgh. Tests of train resistance on South Side Elevated RR of Chicago and automobile tests for competition of *Chicago Times Herald*. Investigated traction problems for Sprague Electric Co., 1898. Private practice in NYC, made electric motor power tests for Brooklyn Elevated RR, Boston Elevated RR and Brooklyn Rapid Transit Company. Reported on power handling of freight at Savannah Terminal of Georgia Central RR Co., designed 1st combined electric hoist and traveler, designed and patented the Lundie Ventilated Rheostat. Consulted for various RRs and other companies in the US. Called to London, England on electrification of the Metropolitan Underground System of RRs. Worked on electric RR work in Canada. Reported to General Electric Co. on conditions and prospects for water power development and utilization of electric power on Isthmus of Panama, 1904. Vice pres. and gen. mgr. of the Panama-American Corp on the Isthmus. Later engaged by the Birmingham Southern RR Corp. in charge of investigating the cost of freight movements on raw material from mines to plants. Efficiency expert outlining plan for location and const. of steel mills for the Tennessee Coal, Iron and RR Co. Designed and patented the Lundie tie-plate for promotion of safety, economy and ease of movement on RRs; and a duplex rail anchor. Formed the Lundie Engineering Corp., 1919, to manufacture and introduce his inventions, the telpher, the Lundie ventilated rheostat, and the Lundie tie-plate. Technical advisor on engr. projects for the United Central America Corp., 1921. Mem: ASCE, Am. Inst. of Consulting Engrs., Am. Assn. for the Advancement of Science, Am. Iron and Steel Inst., Am. Inst. of Electrical Engrs., Am. Railway Engr. Assn., Mason, Knight Templar, Royal Soc. of Arts, Railroad Club, Old Colony Club, Am. Iona Soc., St. Andrew's Soc. and leisure clubs. Refs: TASCE, WWE, WAB, DAB, WWW.

MCCALLUM, DANIEL CRAIG; b. Johnston, Scotland, Jan. 21, 1815; d. Brooklyn, NY, Dec. 27, 1878; w. Mary McCann; c. three. Came with his family to Rochester, NY. Elementary education. Became a carpenter, builder and architect. Originated and patented an arched truss form of bridge. Moved to NY, 1852. Assoc. with Samuel Roberts, const. engr. of the High Bridge over the Harlem River. Gen. superintendent of the NY & Erie RR, 1855-56. Pres. of the McCallum Bridge Co., 1858-59. Consulting engr. for the Atlantic & Great Western RR. Appt. military dir. and supt. of RRs in the US, Feb. 11, 1862. Given war powers to seize and operate all RRs and equipment necessary for

Civil War. Commissioned col. and given position of aide-de-camp on the staff of cmdr.-in-chief. Brevetted maj. gen. on Sept. 24, 1864. Maj. gen., March 13, 1865. Mustered out of service, July 31, 1866. Inspector of the Union Pacific RR. Publ: *McCallum's Inflexible Arched Truss Bridge Explained and Illustrated* (1859), "United States Military Railroads" (1866), *The Water Mill and Other Poems* (1870). Refs: DAB, WWW, WAB, CAB.

CHARLES MACDONALD (1837-1928)
PRESIDENT OF ASCE, 1908-1909

MACDONALD, CHARLES; b. Gananoque, Ontario, Canada, Jan. 26, 1837; d. Gananoque, Ontario, Canada, July 8, 1928; f. William Stone; m. Isabella (Hall); w. Sarah Louise Willard; c. William Stone, Mary Louise, Lillie Paine. Attended public schools and preparatory school of Queen's College, Kingston, Ontario. Worked on surveys for Grand Trunk RRs in Canada, 1852-53. Came to the US in 1854, entered RPI, civ. engr., 1857. Asst. engr. on const. of ext of the Grand Trunk RR from Huron to Detroit, MI, 1858. Lived in Gananoque until 1863. In charge of surveys and const., Philadelphia & Reading RR, Feb. 1863. Became citizen of the US, 1863. Enlisted in 26th Pennsylvania Volunteer Infantry, an emergency regiment raised to meet the invasion of PeA by Southern troops during the Civil War. Captured by the Confederates at the Battle of Gettysburg. Chief engr. of the Reading & Columbia RR and Perkiomen RR, 1867. Moved to NY, 1868. Engaged in iron bridge const. Wrote a pamphlet on the subject. Supervised const. of bridges on extension of the Delaware, Lackawanna & Western RR Co. to Hoboken, NJ. Bridge engr. in NYC, 1868- . Engaged in const. of bridges for some important RRs in the East. Designed and built the Point St. bridge in Providence, RI, 1871-72. Organizer and pres. of the Union Bridge Co., 1884-1900. Supervised design of the Poughkeepsie Bridge over the Hudson River, 1887-88, Hawkesbury Bridge in Australia, 1886-87, and the Sixth St. bridge in Pittsburgh, PA. Asst. in building the Leavenworth Bridge in KS, Memphis Bridge in TN, Winona Bridge in MN, Cairo Bridge in IL, and Merchant's Bridge in MO. Promoter and founder of American Bridge Co., sold to the Union Bridge Co. in 1900. Vice pres. of American Bridge Co., 1900-19s01. Retired from active work in 1901. Acted as consultant on Quebec Bridge over the St. Lawrence River during its reconstruction, 1910. Appt. as trustee of the Brooklyn Bridge. Early promoter of suspension bridge over the Hudson River at 57th St. Prepared plans and a model for a structure in 1896. Proposed use of high carbon steel for eyebars, 1879. Trustee of the Stevens Inst. of Technology. Hon. LL.D. from the Queen's Univ. in Canada. Trustee, RPI. Mem: ASCE (dir., 1871, 1874-76, vice pres., 1893-1894, pres., 1908-1909), Canadian Soc. of Civ. Engrs., Am. Inst. of Mining and Metallurgical Engrs., St. Andrews Soc., Engrs., Lawyers, Univ. and Union Clubs and Century Assn. of NYC, and Engrs. Club of Montreal, Eye Bar Club of NY. Publ: *Journal of a Voyage to Egypt and Italy* (1907). Refs: WWW, WAB, TASCE, RPI Bio.

MACDONALD, THOMAS HARRIS; b. Leadville, CO, July 23, 1891; d. April 7, 1957; f. John; m. Elizabeth (Harris); w. Bess Dunham, March 7, 1907. Moved to Iowa with his family in 1884. Studied at Iowa State College receiving degree in civ. engr. in 1904. Upon graduation took a position as asst. prof. of civ. engr. to perform studies on highways in a newly-created commission to study highway improvements. By 1907, at the age of 26, he was appointed State Highway Engr., and in 1913, with the creation of a three-man State Highway Commission, he was appointed chief engr. During his tenure, IA became one of the 1st states in the Midwest to have a statewide system of main roads. As a result of the attention generated by his work, he became a leader among highway officials and was intimately involved in the efforts of the Am. Assn. of State Highway Officials in efforts to secure federal-aid highway legislation. In 1919 he came to Washington, DC to serve as Chief, Bureau of Public Roads, a position he was to hold through the tenures of 7 different U.S. presidents, until his retirement in 1953. During his tenure as chief of the Bureau of Public Roads, he championed the development and application of quantitative measures in highway location and design, such as traffic studies, measurement of materials properties, economic impacts and many other principles that today are standards. He sponsored the organization of the Highway and Transportation Education Committee (later to become a part of the Highway Research Board) in 1920. Sponsored the organization of the Pan American Highway Congress in 1924, the 1st President's Highway Safety Congress, 1945. He served as a member of the Official Commission in the Alaska Highway, later he was charged with the responsibility for its construction. He strongly opposed a superior administrative role for the transportation agency at the federal level, and promoted the state-federal partnership that guided developments during the pre-interstate era. He received an hon. degree from Iowa State Univ. in 1929 and was made an hon. mem. of ASCE in 1943; following his retirement from federal service, he headed the Highway Research Center at Texas A&M Univ. He authored many papers on highway engineering and administration. Refs: HN.

MCCLELLAN, CARSWELL; b. Philadelphia, PA, Dec. 3, 1835; d. St. Paul, MN, March 6, 1892; f. Samuel; m. Margaret Carswell (Ely); w. Annis Davis. Matriculated at the Univ. of Pennsylvania, class of 1854. Left in 1851 to complete his college course at Williams, grad. A.B. 1855, A.M. 1858. Enlisted in Volunteer Army as ensign in the 32nd NY Regiment, May 1862. Wounded at the battle of Malvern Hill. Topographical asst. to the adjutant gen., and special aide on the staff of Gen. Andrew A. Humphreys, July 3, 1862. Participated in battles at Fredericksburg and Chancellorsville, VA. Wounded at Gettysburg. At the headquarters of the Army of the Potomac until April 1864. Asst. adjutant gen. 1st Division Calvary Corps, Army of the Potomac and of the 3rd Division, 5th Army Corps, April-Aug.1864. Taken prisoner in the battle of Weldon RR, Aug.19, 1864. Parolled Nov. 16, 1864 when he resigned. Engr. in charge of location and const. works on St. Louis, Vandalia & Terra Haute, Northern Pacific, St. Paul & Pacific and other western RRs, 1867-81. US asst. civ. engr., 1881-92. Publ: *Personal Memoirs and Military History of Ulysses S. Grant vs. the Record of the Army of the Potomac* (1887). Refs: Lamb, WWW, CAB, BDNA, WAB.

MCMATH, ROBERT EMMET; b. Varick, NY, April 28, 1833; d. Webster Groves, MO, May 31, 1918; f. Alla; m. Elizabeth Parshall (Homan); w. Frances Brodie; w. 2nd Eleanor C. Trent; c. Frances Isabella, Thomas Brodie, Robert Homan and Frances Charles. Early education in academies of Waterloo and Genesco, NY. Grad. from Williams College with degree of A.B., 1857. Moved to St. Louis, MO, Oct. 1858. Engaged in surveys, designs and const. to improve the Mississippi River and some of its tributaries. Deputy county surveyor of St. Louis, 1859-62. Asst. engr. in US Coast Survey, 1862. Worked in defense preparations in Philadelphia against Confederate raiders, June 1863. Topographical survey of Arlington Heights (Arlington National Cemetery) in Washington, DC, 1864. Mapped territory in FL between the St. Johns River and the Atlantic. Mapped the Sheepscott River in Maine from Bath to the ocean. Surveys of San Juan River and Greytown Harbor in connection with an interoceanic canal proposed by Cental American Transit Co. with a grant from the Nicaraguan govt., Feb. 1865. Accepted position of asst. engr. in the US Army Corps of Engr. Corps, fall 1865-1883. On hydraulic survey of the Illinois River, 1867-68. Asst. engr. on survey of the Arkansas River, 1869. In charge of improvement of the Illinois River, Sept. 1869-July 1872. In charge of improvement of harbors of St. Louis, MO and Alton, IL, Aug. 1872-April 1873. Principal civilian asst. on improvement of the Mississippi River, April 1873-78. Employed by Mississippi River Commission, 1880-83. Developed improved methods for deepening channels and preventing flood damage. Appt. sewer commissioner of St.Louis, 1883-91. Devised McMath formula to help determine proper size for storm water sewers. Private practice of engr. as consulting engr. in St. Louis, 1891-93. Elected pres., bd. of public improvements of St. Louis, April 1893-1901. In practice as consulting engr., 1901-May 31, 1918. Dir., Canadian Bridge Co. Mem: ASCE (dir., 1889), Engrs. Club of St. Louis (vice pres., 1895, pres., 1886), Am. Soc. for Municipal Improvements (pres., 1901), Sons of the Am. Revolution. Publ: "Determination of the Size of Sewers," and other professional papers. Refs: DAB, WAB, TASCE.

MCMILLAN, CHARLES; b. Moscow, Russia, March 24, 1841; d. Sept. 19, 1927; f. Alexander; m. Elizabeth (Platt). Educated at Protestant Chapel in Moscow until 1854 and in Hamilton, Ontario, Canada, 1855-56. Entered RPI in Sept. 1856, grad. with degree of civ. engr., 1860. Asst. engr. Brooklyn Water Works, 1860. Draughtsman, Croton Water Works, NY, 1861-63. Asst. engr. in charge of pipe distribution and rates of engrs., 1863-64. Asst. engr. in const. of reservoir in Central Park, NY, 1864. Civ. engr., McMillan & Gould, Titusville, PA, 1865. Prof. of geodesy and road engr., RPI, 1865-71. Prof. of civ. and mech. engr., Lehigh Univ., 1871-75. Prof. of civ. engr. and applied mathematics, Princeton Univ., 1875-1914. Consulting engr. in Troy, NY, Bethlehem, PA and Princeton, NJ. Publ: Editor, *Smith's Topographical Drawing* (1885-) . : WWW, RPI Bio, CAB.

MAHONE, WILLIAM; b. near Monroe, Southampton County, VA, Dec. 1, 1826; d. Washington, DC, Oct. 8, 1895; f. Fielding Jordan; m. Martha (Drew); w. Ortelia Butler; c. two sons and one daughter. Early education under his father, attended school for two years. Grad. from Virginia Military Inst., 1847. Taught at Rappahannock Military Academy for two years. Engr. of Orange & Alexandria road building project. Civ. engr. and constructor, Norfolk & Petersburg RR, 1851, pres., chief engr. and supt., 1861. Joined the Confederate Army, 1861, commissioned lt. col. of Virginia Volunteers, became col. of 6th Virginia Regiment of Eastern Volunteers, took part in capture of the Norfolk Navy Yard in April, 1861. Commanded Norfolk District until its evacuation, May 1862. Drewry's Bluff defenses of the James River. Promoted to the command of the 2nd Brigade, Huger's Division, Magruder's Command, took part in battles of Seven Pines and Malvern Hill. Commanded the 3rd Brigade, R. H. Anderson's Division, 1st Corps, Army of Northern Virginia in the Chancellorsville Campaign. Promoted brig. gen. in April 1864, maj. gen. in Aug. 1864. Took part in battle of Petersburg, July 1-3, 1864. Mem. of the North Carolina Senate, 1863-65. Pres. of Atlantic, Mississippi & Ohio RR, 1867-73. Elected pres. and rebuilt the Norfolk & Tennessee RR. Organized and became prominent leader of the Readjuster Party, favored partial repudiation of the state debt and secured control of the state legislature. Elected US Senator, 1880-March 3, 1887. Defeated for re-election, 1886 by J. W. Daniel. Refs: BDNA, CAB, Lamb, DAB, WWW, WAB.

MAIN, CHARLES THOMAS; b. Marblehead, MA, Feb. 16, 1856; d. Winchester, MA, March 6, 1943; f. Thomas; m. Cordelia Green (Reed); Elizabeth F. Appleton; c. Charles Reed, Alice Appleton, and Theodore. Early education in the Marblehead schools. Entered MIT, grad. in 1876 with degree in mech. engr. Asst. instructor in mech. engr. dept. at Massachustts Inst. of Technology while doing advanced work, 1876-79. Draftsman in textile mills at Manchester, NH, fall 1879-Jan. 1, 1881. Engr. with Lower Pacific Mills, Lawrence, MA, 1881-87, asst. supt., March 1886, supt., July 1887-Dec. 31, 1891. Rebuilt the plant, reorganized machinery and installed a new power plant. Three years as alderman of Lawrence, 1887-89, trustee public library and mem. of school committee. Engr. and mill work in Providence, RI, 1892-93. Formed partnership in Boston with Francis Winthrop Dean, Jan. 1,1893-Jan. 1, 1907. Organized firm of Charles T. Main, 1907, incorporated in 1926. Planned and const. cotton mills in New England and the South, advised on building of steam and water power plants across the country. Chair of Water and Sewer bd. in

Winchester, MA, 1894-1905. Moved to Winchester, 1896. Supervised const. of new plants for American Woolen Co., including the Wood Worsted Mills in Lawrence, MA, 1899-1924. Designed and supervised const. of a municipal lighting plant in Marblehead and a steam electric plant for the Lynn, MA Gas & Electric Co. Steam and waterpower projects included the Conowingo Dam across the Susquehanna River in MD and the Keokuk Dam across the Mississippi. Firm designed almost 80 hydroelectric plants. Consulting engr. during World War I to the Const. Division, Quartermaster Corps of the War Dept. One of nine Am. engrs. sent to France to consult with French authorities on reconstruction work, Dec. 1918. Charles T. Main Award est. in 1919, given annually to student mem. of the Am. Soc. of Mech. Engrs. Served on the Am. Engr. Council. Delegate to the World Power Conference held in Tokyo, Japan, 1929. Drafted the 1st code of ethics adopted by any engr. soc. in the US while pres. of the Boston Soc. of Civ. Engrs. Assn. Medal of National Assn. of Cotton Manufacturers, 1911, Desmond Fitzgerald Medal of the Boston Soc. of Civ. Engrs., 1913, and Gold Medal of the Am. Soc. of Mech. Engrs., 1935. Hon. degree from Northeastern Univ., 1935. Mem: ASCE, Boston Soc. of Civ. Engrs. (pres., 1912), Engrs. Club of Boston (pres., 1914-25), Am. Inst. of Consulting Engrs. (pres., 1929), Am. Soc. of Mech. Engrs. (pres., 1918), MIT Alumni Assn. (pres., 1901), the Am. Assn. for the Advancement of Science, Newcomen Soc. Publ: *Notes on Mill Construction* (1886), published several papers on methods for evaluating water power plants. Refs: DAB, WWE, WWW, WAB.

MARSHALL, WILLIAM LOUIS; b. Washington, KY, June 11, 1846; d. Washington, DC, July 2, 1920; f. Charles Alexander; m. Phoebe Ann (Paxton); w. Elizabeth Hill Colquitt; c. one daughter. Grammar school at Kenyon College, 1859-60, then entered the college. At start of the Civil War, enlisted in the 10th Kentucky Cavalry, Aug. 16, 1862-Sept. 17, 1863. Appt. as cadet to the US Military Academy June 1864, grad. June 1868. Assigned 2nd lt. to the Corps of Engrs. Served with batallion of Engrs. at Willet's Point, NY, 1868-70. Asst. prof. of natural and experimental philosophy at the US Military Academy, 1870-71. Asst. to Lt. G. M. Wheeler in exploration of the Rocky Mountain region, 1872-76. Discovered Marshall Pass, 1873, and gold placers in the Marshall Basin of the San Miguel River, CO, 1875. Asst. engr. on river improvement projects in AL, GA and TN, 1876-81. In charge of improvement of section of the Mississippi River and const. of levees in MS, LA and AR, 1881-84. Worked on improvement of the Fox and Wisconsin Rivers and improvement of harbors in the Milwaukee District, 1884-88. Promoted to capt., 1882, maj. in 1885. Worked on govt. engr. improvements in the Chicago area, including charge of the Chicago and Calumet Rivers and harbors, const. Hennepin Canal, connecting Illinois River at Lasalle with the Mississippi River at Rock Island, 1888-99. Patented improvements connected with it: a combined breakwater and beach, May 12, 1890, automatic movable dam or sluiceway gate, March 23, 1897, and an automatic dam, weir or gate, Dec. 28, 1897 and Jan. 4, 1898. On bd. to advise on water supply of Washington, DC (pres.), Missouri River Commission, and Lincoln Park Bd., Chicago. Also consulting engr. for the Lincoln Park Bd. In charge of fortification and river and harbor work in NYC, 1899. Completed the Ambrose Channel, planned and completed extension of Governor's Island and const. coast defenses. Commissioned col., Aug. 27, 1907. Commissioned Chief of Engrs. with the rank of brig. gen., July 2, 1908. Retired from active service, June 11, 1910. Served on bd. to report on necessary defenses of the Panama

Canal, 1909. Appt. consulting engr. to the Secy. of the Interior. Served on several bds. dealing with projects of the US Reclamation Service. Reported on possible hydroelectric power development projects in different parts of the country. Refs: WAB, DAB.

MASON, CLAIRBOURNE RICE; b. Chesterfield County, VA, Nov. 28, 1800; d. Swop's Station, VA, Jan. 12, 1885; f. Peter Mason; m. Elizabeth; w. Drucilla W. Boxley; c. eleven. Ran away from home then he was eight, worked on a farm in PA. Carried mail in MD. Worked in Washington, DC as an apprentice to a ship's carpenter. Contractor with the const. of the Midlothian RR of VA, 1829. Began the Louisa RR, later the Chesapeake & Ohio RR, acted as supt. Contractor on Virginia Central RR. Raised a co. of Confederate volunteers at the start of the Civil War, chosen capt. Service under Andrew "Stonewall" Jackson. Returned to RR contracting after the war. Contracts included: Chesapeake & Ohio RR, Valley RR of Virginia, Baltimore & Ohio RR, Cincinnati Southern RR, Kentucky Central RR, Richmond, Fredericksburg & Potomac RR, Richmond & Allegheny RR, Richmond & Mecklenburg RR, Kentucky Union RR, Virginia & North Carolina RR Extension and Southern Pennsylvania RR. Refs: DAB.

MASON, WILLIAM PITT; b, NYC, Oct. 12, 1853; d. Little Boars Head, NH, Jan. 25, 1937; f. James; m. Emma (Wheatley); w. Emilie Eliza Harding; w. 2nd Margaret Delevan Betts; c. George Harding, William Pitt. Early education at the Bernard French Inst., NYC. Entered RPI Sept. 1871, grad. 1874 with degree of civ. engr. Studied for a year in Europe and in chemical laboratory of Harvard. Asst. in chemistry and natural science, RPI, Sept.1875. Received B.S. there in 1877. Attended Albany Medical College at the same time, received M.D., March 1881. Asst. prof. of chemistry and natural science in 1882, prof. of analytical chemistry in 1885. Studied cholera epidemic in Messina, Sicily and water supply systems of other European cities, 1887. Studied under Louis Pasteur in Paris in 1889 and 1893. Also took courses at the Sorbonne, the Ecole Centrale des Arts et Manufactures and the Ecole Polytechnic in Paris, studied water supply systems of London, Glasgow, Paris, Vienna, Rome, Genoa, Florence and Turin and the sewage farms outside of Paris, 1893. Studied plankton and other material which caused bad taste and smell in water with Otto Zacharias at Plon, Germany, summer 1908. Expert on sanitary engr. and subject of water supply. Served as consultant to cities and private institutions throughout the country, trained a large number of sanitary engrs. Mem. of commission for the revision of the US Pharmacopoeia, 1890 and the US Assay Commission, 1896. Pres. of the Hygiene Division of the 8th Intl. Congress of Applied Chemistry, Washington DC, 1912. Edited *Notes on Qualitative Analysis*, Sept. 1882. Honarary LL.D. from Lafayette College, 1908, and Sc.D. from Union Univ., 1917. Mem: ASCE, Am. Assn. for the Advancement of Science, Am. Philosophical Soc., Am. Chemical Soc, Am. Water Works Soc., (pres., 1909), New England Water Works Soc., Am. Public Health Assn., Am. Inst.of Consulting Engrs., Franklin Inst., Royal Sanitary Inst. of Great Britain, Washington Academy of Sciences, Am. Inst. of Chemical Engrs., and Assn. Generale des Ingenieurs, Architects et Hygenistes Municipaux of France, Belgium, Switzerland and Luxembourg (hon. mem.). Publ: *Water Supply* (1896, 1918), *Examination of Water* (1899, 1931), and many technical papers. Refs: RPI Bio, WAB.

MAYNARD, GEORGE WILLIAM; b. Brooklyn, NY, June 12, 1839; d. Boston, MA, Feb. 12, 1913; f. George Washington; m. Caroline Augusta (Eaton); w. Fannie Atkin; c. one daughter. Entered Columbia College in 1855, grad. in 1859. Asst. to the prof. of chemistry. Studied abroad at Gottingen, specialized in chemistry, physics and mineralogy, 1860. Later went to School of Mines at Clausthal to study mining and mineralogy. Devised process for the treatment of pyritic ores at Wicklow, Ireland, 1863-64. Returned to the US. Opened engr. office and chemical laboratory under the firm name of Maynard & Tiemann. Degree of A.M. from Columbia College, 1864. Est. an engr. and assay office in Gilpin County, CO for three years. Returned east in 1867, took charge of small plant for manufacturing sulphuric acid on Staten Island. Accepted professorship of metallurgy and practical mining at RPI, 1868. Returned to New York a few months later. Went to England to negotiate sale of an iron property in the southern states, 1873. Opened an office in London, became consulting engr. for sundry steel works in England and Wales. Directed 1st large scale tests in England of the Bessemer process. Remained abroad for six years, erected a copper smelting plant in Russia for a British company. Returned to the US, sold the Am. rights to the Bessemer process. Spent the rest of his life as a consulting engr. in NYC. Worked in Nova Scotia, Newfoundland, British Columbia, the Yukon, Mexico, and Cuba. Mem: Am. Inst. of Mining Engrs. (vice pres.), New York Academy of Sciences, Am. Geographical Soc., Am. Soc. of Mech. Engrs., Long Island Historical Soc. and Iron and Steel Inst. of Great Britain. Published several papers in technical publications. Refs: DAB, CAB, RPI Bio.

MEADE, GEORGE GORDON; b. Cadiz, Spain, Dec. 31, 1815; d. Philadelphia, PA, Nov. 6, 1872; f. Richard Worsam; m. Margaret Coates (Butler); w. Margaretta Sergeant; c. four sons including John Sergeant and George, and two daughters. Attended Mount Airy School near Philadelphia, withdrawn after his father died. Sent to school of Salmon P. Chase in Washington, DC. Attended Mt. Hope School in Baltimore. Received appointment on his second application, became a cadet at the US Military Academy, Sept. 1, 1831. Graduated 19th among 56 members of the class of 1835. Asst. surveyor of Long Island RR. Brevet 2nd lt. of the 3rd Artillery, Dec. 31, 1835. Ordered to FL in war against the Seminoles, 1835-36. Left as a result of illness. Ordered to Watertown Arsenal, MA on ordnance work, 1836. Resigned from Army, Oct. 26, 1836. Asst. civ. engr. of Alabama, Florida & Georgia RR, 1836-April 1837. Principal asst. engr. on a survey of improvement of mouths of the Mississippi for navigation, April 1837-Feb. 1839. Asst. to the joint commission for est. a boundary between the US and Texas, 1840. Returned to Washington, DC the same year. Civilian asst. of survey of the northeastern boundary, Aug. 1840. Reinstated in Army, May 19, 1842, appt. 2nd lt. in Corps of Topographical Engrs. Continued on the northeastern boundary survey until Nov. 1843. Designed and const. lighthouses in the Delaware Bay in Philadelphia, 1843-45. Ordered to Aransay Bay, TX with Taylor's Army of Occupation, Aug. 1845. During the Mexican War, engaged in battles of Palo Alto and Resaca de la Palma, May 1846. Brevetted 1st lt. on Sept. 23, 1846 for reconnaissances performed at Monterey. Participated in siege of Veracruz, May 9-29, 1847. On return to Philadelphia, presented with sword for his services. Const. lighthouses in Delaware Bay, made surveys and maps of FL reefs, 1847-

49. In active service against Seminoles, 1849-50. On lighthouse duty in Delaware Bay, 1850-51. Promoted 1st lt. in Corps of Topographical Engrs., Aug. 4, 1851. Worked at the Iron Screw Pile Lighthouse on Carysfort Reef in FL, 1851-52. At Sand Key, 1852-56. Capt., Corps of Topographical Engrs., May 17, 1856. Geodetic survey of the Great Lakes, Detroit, MI, 1856. In charge of the Northern Lake Surveys, 1857-61. Brig. gen. of Volunteers at the start of the Civil War, Aug. 31, 1861. Assigned to command of the 2nd Brigade of Pennsylvania Reserves in the Army of the Potomac, 1861-62. 1st active service in the defenses of Washington, DC, asst. in const. of Ft. Pennsylvania. Transferred to McDowell's Army, March 1862. After evacuation of Manassas, went to the Dept. of the Shenandoah. Ordered to the peninsula under Gen. McClellan, June 1862. Promoted to maj., Corps of Topographical Engrs. of the regular Army, June 18, 1862. Took part in battles of Mechanicsville, Gaine's Mill and Glendale. Sick leave from wound at Glendale, July-Aug. 1862. Participated in Second Bull Run, Aug. 29-30, 1862. Placed in temporary command at South Mountain, Sept. 14, 1862. Placed in temporary command of the I Corps, led for remainder of the battle. Under Gen. McClellan in the pursuit of Lee to Falmouth, VA, Oct.-Nov. 1862. Maj. gen. of Volunteers, Nov. 29, 1862. Regular command of the V Corps, Dec. 25, 1862. Placed in command of the Center Grand Division, composed of the III and VI Corps, Jan. 26, 1863. In command of the V Corps, Feb. 5. In the Battle of Chancellorsville, May 24, 1863. In command of the Army of the Potomac, June 28, 1863. Directed Battle of Gettysburg, July 1-3, 1863. After battle at Gettysburg, received thanks of Congress, Jan. 28, 1864. Promoted brig. gen. of the regular army, July 3, 1863. Powers cut back when Grant decided to accompany the main army in Virginia. Retained in command of the Army of the Potomac until Appomattox. Promoted maj. gen. in regular army, Aug. 18, 1864. Successively placed in command of Military Division of the Atlantic and Dept. of the East with headquarters at Philadelphia, Aug. 1866-Jan. 1868. Transferred to Atlanta, GA, in command of 3rd Military District of Dept. of the South (GA, AL and FL), Jan-Aug. 1868. Served with dept. comprising same states with SC and FL, Aug. 1868-March 12, 1869. Transferred to command of Military Division of the Atlantic with headquarters in Philadelphia. Commissioner of Fairmount Park, Philadelphia, 1866 until his death. Received degree of LL.D. from Harvard in 1865. Buried with military honors. Equestrian statue of Gen. Meade dedicated in Fairmount Park, Philadelphia on Oct. 18, 1887. Mem: Am. Philosophical Soc., Pennsylvania Historical Soc., Philadelphia Academy of Natural Sciences. Refs: Lamb, DAB, CAB, BDNA, DR, WAB.

MERRILL, WILLIAM EMERY; b. Ft. Howard, Green Bay, WI, Oct. 11, 1837; d. Cincinnati, OH, Dec. 14, 1891; f. Moses E.; m. Virginia (Slaughter); w. Margaret E. Spencer; c. eight children including two sons. Appt. to US Military Academy, June 1854. Grad. 1st in his class in 1859, assigned to Corps of Engrs. Asst. prof. of engr. at the US Military Academy for almost a year. Served as military engr. during Civil War in the Dept. of the Ohio, the Army of the Potomac, March-April 1862. Chief engr. of the Army of Kentucky, Oct. 12, 1862-May 25, 1863. Chief engr. of the Army of the Cumberland, Jan. 27, 1864-June 27, 1865. Captured during McClellan's campaign in WV, Sept. 12, 1861. Kept as prisoner until the following Feb. Escaped in Nov. but was recaptured. Wounded near Yorktown, VA, April 1862. Brevetted capt. for gallantry. Engaged in the Cedar Mountain and Manassas Campaign, transferred to the

West to fortify Covington and Newport, Sept.-Oct. 1862. Capt., March 3, 1863. Under Rosecrans in the Chickamauga Campaign, under Thomas in the battle of Missionary Ridge and under Sherman in the advance on Atlanta. Const. fortifications for protection of RRs supplying Sherman's army. Brevetted maj., March 7, 1867, lt col., Feb. 20, 1883, and col. for merits in battles of Chickamauga, Lookout Mountain, Missionary Ridge, Resaca and New Hope Church. Chief engr. of Division of the Missouri for three years until 1870. The rest of his career on river and harbor improvement work of the Corps of Engrs. Originated canal on the Ohio River from Pittsburgh to its mouth. Charged with improvement of this river in 1870. Sent to Europe at his own request in 1878 to study improvement of non-tidal rivers through locks and movable dams. On return to US, secured from Congress an appropriation for the Davis Island Lock and Dam (National Historic Civil Engineering Landmark, 1985) below Pittsburgh, completed in 1885. Project extended to the entire river, not completed until 1929, after his death. US representative at the Congress of Engrs. in Paris, 1889. Mem: ASCE (dir., 1883), Engrs. Club of Cincinnati, Ohio (organizer). Publ: *Iron Truss Bridges for Railroads* (1870), and later studies of the improvement of non-tidal rivers and of inland navigation in France and the US. Refs: DAB, CAB, WAB.

COLONEL WILLIAM E. MERRILL (1837-1891)

MILLER, EZRA; b. near Pleasant Valley, NJ, May 12, 1812; d. Mahwah, NJ, July 9, 1885; f. Ezra Wilson; m. Hannah (Ryerson); w. Amanda J. Miller; c. three sons and two daughters. Educated in schools in NYC, Rhinebeck and Flushing, Long Island. Educated himself in topographical, mech., civ. and hydraulic engr., practiced profession in and about NY. Engaged in military studies, active in the state militia. Enlisted in co. of artillery of the 2nd NY

Militia, 1833. Adjutant in 1839, lt. col., 1840 and col., 1842 then retired. Settled at Ft. Hamilton, NY, 1841. Practiced profession until 1848. Moved to Rock County, WI to survey public lands, 1848. Elected Justice of the Peace for two terms. Engaged in RR survey and const. work. Commissioned col. in the Wisconsin Militia, 1851. Mem. of the WI Senate for one term, 1852. Studied improvement of coupling RR cars. Patented a car coupler on March 31, 1863. Patented improved RR car platform, coupler and buffer, Jan. 31, 1865. Returned to Brooklyn NY, 1867. Purchased farm in Mahwah, NJ, raised livestock. Elected to state senate in NJ, 1883. Refs: WAB, CAB, DAB.

MILLINGTON, JOHN; b. Hammersmith, England, May 11, 1779; d. Richmond, VA, July 10, 1868; f. Thomas Charles Millington; m. Ruth Hill; w. Emily (Hamilton); c. _____ ; w. Sarah Ann (Letts); c. _____ . Financial difficulties forced him to forego studies at the Univ. of Oxford. He turned his attention to the study of law, being accepted by the legal brotherhood in 1803. No evidence of any serious attention being given by him to the practice of law although for several years he carried on a considerable practice as a patent agent. Turning from law he gave himself entirely to engr. He was engr. of the West Middlesex Water Works. Also served as supt. engr. of the Royal Grounds in London. Elected to the Royal Soc. of Arts. From 1921-25 he served as Steward of the Soci. Lectured on Natural Philosophy in the Royal Inst. until 1829. July 7, 1817, he was elected prof. of mech. in the Inst. without salary and served until 1829. Elected one of the original members of the Royal Astronomical Soc., Feb. 29, 1820; served as secy. from 1823-25. Received life mem. in the Soc. in 1825. Dec. 2, 1823, the Linnean Soc. of London elected him a fellow. Became vice pres. of the Mechanics Inst. Served on the 1st faculty on the Univ. of London. About 1829 or 1830 he left England for US to serve as Chief Engr. of Mines and Supt. of a Mint for an English co. After his 1st wife died he began an extended tour of the U.S. Fell seriously ill in Philadelphia and upon his recovery married a Miss Letts. Opened a scientific shop for the supply of all the various machines, instruments, apparatus and materials required for "mechanical, philosophical, mathematical, optical and chemical purposes." In 1835 he was elected to the chair of chemistry, natural philosophy and engr. in the College of William and Mary. In July 1848 (the year the college of William and Mary was closed because of student unrest) he was elected to the chair of natural science at the newly organized Univ. of Mississippi. On July 13, 1853, he resigned and accepted the chair of chemistry and toxicology at the Memphis Medical College. In 1859 or 1860 he relinquished his chair in the Memphis Medical College and settled in the town of La Grange, TN. Civil War broke out and he moved to Philadelphia until the end of the War. Returned to Richmond where he spent his remaining days with his eldest daughter. Died July 10, 1868, and is buried in Bruton Parish Churchyard, Williamsburg, VA. Publ: *Epitome of the Elementary Principles of Mechanical Philosophy* (London, 1823; 2nd edition, 1830), *Elements of Civil Engineering; Being an Attempt to Consolidate the Principles of Various Operations of the Civil Engineer into One Point of View*, 8 vols., 273 illustrations (Philadelphia, 1839); *Engineering and the Application of Mechanical Philosophy to the Arts*. Refs: HN.

MILNER, JOHN TURNER; b. Pike County, GA, Sept. 29, 1826; d. Newcastle, AL, Aug. 18, 1898; f. Willis Jay; m. Elizabeth (Turner); w. Flora J. Caldwell; c. Henry Willis, Elizabeth Cauldwell, Lilian, and Florence. Grew up on a farm, simple schooling, worked in his father's gold mines in Lumpkin County. Practical education in RR const. Attended Univ. of Georgia, 1843, left at end of his third year because of illness. Principal asst. in const. of the Macon & Western RR. Asst. engr. in const. of Muscogee RR. Joined the gold rush in 1842, drove a team of oxen to OR and CA. Became city surveyor of San Jose. Returned east, 1852. 1st asst. engr. of the Alabama and Florida RR. Appt. to plan RR connecting the navigable parts of the Alabama and Tennessee Rivers to transport mineral resources. Const. suspended during the Civil War. Chief engr. of South and North Alabama RR eventually built following the war, completed Oct. 1, 1872. Est. the Oxmoor furnaces on a section of iron ore sites near Elyton, AL in 1862, which served as a Confederate Army munitions plant. In building the South and North Alabama RR, bought sections of land with intention of founding industrial city at the point where his RR and the Alabama & Chattanooga would intersect. Formed Elyton Land Co. Dec. 19, 1870, which founded Birmingham in 1871. Organized the Newcastle Coal & Iron Co., 1873. Est. Milner Coal & RR Co., 1879-89. State senator from Jefferson County, 1888-96. Publ: "Report of the Chief Engineer of the South & North Alabama RR Co." (1859). Refs: DAB, WAB.

MOORE, ROBERT; b. New Castle, PA, June 19, 1838; d. July 24, 1922; f. Henry C.; m. Amelia (Whippo); w. Alice Filley; c. Charles Whippo Moore. Early engr. training as flagman and rodman under his father, a civ. engr. in RR dev in IN and OH. Educated in public schools. Grad. from Miami Univ. of Ohio, A.B., June 20, 1858, and A.M. 1866. Civilian asst. engr. with US Army Corps of Engr. at Camp Nelson, KY. Chief engr. of Terre Haute and Indianapolis RR, 1868-69. Built road from Belleville to Duquoin, IL, 1869-70. Asst. engr. in completion of RR from Pleasant Hills, MO to Lawrence, KS. Chief engr. of road from Lorain to Urichsville, OH, now part of Cleveland, Lorain & Wheeling RR, 1872-73. Located eastern half of Indiana, Decateur & Western RR. Mem. of 1st Bd. of Public Improvements in St. Louis MO, 1877. Sewer Commissioner, 1877-81. Consulting engr., 1881-1920. In charge of reconstruction of the RR floor system of Eads Bridge with St. Louis & Illinois Bridge Co. Consulting engr. of St. Paul and Duluth RR. Chief engr. of St. Louis Merchants Bridge Terminal RR. Consulting engr. for Reorganization Committees of St. Louis Southwestern RR, Rio Grande Western RR and Santa Fe System. Chief engr. of St. Louis, Peoria and Northern RR, a line extending from St. Louis to Peoria, IL, 1894-98. Represented Los Angeles Terminal RR in San Pedro-Santa Monica Harbor controversy. Chief engr. of Wabash, Chester & Western RR and MO extension of Missouri & Illinois RR. Mem. of Brazos River Bd, 1897. Mem. of Bd. of Education in St. Louis, 1897-1913, pres., 1906 and 1910. Mem. of Reform Bd. of Public Schools. Mem. of Southwest Pass Bd., 1899-1900. Special consulting engr. on Missouri and Mississippi River Bridges for the Chicago, Burlington & Quincy RR, 1900-20. Mem: ASCE (dir., 1892-93, vice pres., 1888-1900, pres., 1902), Inst. of Civ. Engrs. of Great Britain, Engrs. Club of St. Louis (pres.), Ethical Soc. (chair, 1891). Refs: TASCE, WWW, WAB.

MORELL, GEORGE WEBB; b. Cooperstown, NY, Jan. 8, 1815; d. Scarborough, NY, Feb. 12, 1883; f. George; m. Maria (Webb); w. Catherine Schermerhorn Creighton; c. none. Grad. from US Military Academy in 1835, 1st in his class. Asst. engr. with Corps of Engrs. in improvement of Lake Erie Harbors. Promoted 2nd lt., Oct. 31, 1826. On OH and MI boundary survey and in const. of Ft. Adams, Newport, RI. Resigned, June 30, 1837. Asst. engr. of const. for Charleston & Cincinnati RR, 1837. With Michigan Central RR 1838-40. Moved to NYC and studied law. Admitted to the bar, 1842, practiced law until 1861. Maj. and division engr. of 1st Division of the NY militia in 1849-52. Promoted to col., 1852-61. Commissioner of US Circuit Court for the Southern District of NY, 1854-61. Appt. inspector, 1st Division of the NY Militia, April 15, 1861. Appt. col., served as quartermaster and chief of staff to Maj. Gen. Sanford, NY Volunteers. Organized troops and sent them to war, April-May 1861. Engaged in defenses of Washington, DC, May 20-July 7, 1861. Involved in operations near Harper's Ferry, VA, July 7-Aug. 21, 1861. Promoted to brig. gen. of US Volunteers, Aug. 9, 1861. Assigned to Army of Potomac. Guarded approaches to Washington, DC, Aug. 21, 1861-March 10, 1862. In the VA Peninsular Campaign, March-Aug. 1862. Battles included, Howard's Bridge, April 4, 1862, siege of Yorktown, April 5-May 4, 1862, capture of Hanover Court House, May 27, 1862, Mechanicsville, June 26, 1862, Gaine's Mill, June 27, 1862, and Malverne Hill, July 1, 1862. Served in campaign of Northern VA and the Maryland campaign, Aug.-Oct., 1862. Engaged in Manassas, Aug. 30, 1862 and Antietam, Sept. 17, 1862. Commanded forces guarding the upper Potomac, Oct. 30-Dec. 16, 1862. Placed in command of draft rendezvous in Indianapolis, IN, Dec. 15, 1863-Aug. 29, 1864. Mustered out of service, Dec. 15, 1864. Settled in Scarborough, NY until his death. Mem: Soc. of the Cincinnati, Union Club, Soc. of the Army of the Potomac. Refs: DAB, WWW, Lamb, BDNA, WAB, CAB.

MORRIS, THOMAS ARMSTRONG; b. Nicholas County, KY, Dec. 26, 1811; d. San Diego, CA, March 22, 1904; f. Morris; m. Rachel (Morris); w. Elizabeth Rachel Irwin; c. five including Thomas O., Eleanora Irwin, Milton A. Moved with his family to Indianapolis in 1821. Worked with a printer, 1823-26. Entered private school of Elizabeth Sharpe, 1826-30. Entered the US Military Academy on July 1, 1830. Grad. 4th in his class, 1834. Commissioned brevet 2nd lt. of artillery, commissioned 2nd lt., Feb. 25, 1835. Resigned the next year. In charge of const. of Indianapolis section of Central Canal in IN. Built "state ditch" saving the city from floods. Chief engr. Madison & Indianapolis RR, supt. building from Vernon to Indianapolis, 1841-47. Chief engr. of Terre Haute & Richmond RR and of Indianapolis & Bellefontaine, OH, 1847-52. Prepared estimates and reports on Peru & Indianapolis RR. Chief engr. of Indianapolis & Bellefontaine, 1852-54, pres., 1854-57. Planned and built union depot at Indianapolis, 1853. Pres., Indianapolis & Bellefontaine RR, 1857-59. Chief engr. of the Indianapolis & Cincinnati, 1859-61. Appt. state quartermaster gen., 1861. Commissioned brig. gen., April 27, 1861. May 1861, ordered to the western part of MA. Mustered in volunteer regiments of western MA unionists. Drove Confederate troops back from Philippi on June 3. Started driving them out of western VA. Reinforcements were refused on July 3. Pursuit action at Carrick's Ford, July 13. Honorably mustered out on July 27, 1861. Declined commission of brig. gen. in Sept. 1862, and as jr. maj. gen. in Oct. Chief engr. of Indianapolis & Cincinnati RR, built Lawrenceburg-Cincinnati section, 1862-66.

Pres. and chief engr. of the Indianapolis & St. Louis, 1866-69. Const. road between Terre Haute and Indianapolis. Receiver of Indianapolis & Lafayette RR, 1869-72. Commissioner to select plans and supt. const. of new state Capitol, 1877. Pres. of the Indianapolis Water Co., 1888. Life Trustee Consumer's Gas Trust Co. Refs: DAB, WWW, WAB.

MORSE, CHARLES ADELBERT; b. Bangor, ME, Jan. 1, 1859; f. Charles B.; m. Elsie (Emery). Educated at the Univ. of Maine, grad. with degree of civ. engr., 1879. Instrumentman, office man and division engr., C. B. & Q. RR, 1880-81. Division engr. with Mexican Central RR, 1881-84. Employed by the C. B. & Q. RR, 1884-85. Transitman, division engr., and resident engr., Atchinson Topeka and Santa Fe RR, Ft. Madison, IA and Pueblo, CO, 1886-1901. Asst. to chief engr., Topeka, KS, 1901-1902. Principal asst. engr., La Junta, 1902-1903. Engr. Eastern Grand Division, Topeka, March-July 1903. Acting chief engr., Topeka, 1903-1904. Asst. chief engr., Topeka, 1904-1905. Acting chief engr., Atchinson, Topeka and Santa Fe RR coast lines, Los Angeles, 1905-1906. Chief engr. Atchinson, Topeka and Santa Fe, RR, Topeka, 1906-1909. Chief engr. Atchinson, Topeka and Santa Fe System, Topeka, 1909-13. Chief engr., Rock Island Lines, Chicago, April 1, 1913-Sept. 1, 1918, June 1, 1919-1929. Asst. dir. engr. and maintenance, US RR Administration, Washington, DC, Sept. 1, 1918-June 1, 1919. Appt. mem. of Bd. of Review, Const. Dept., US Army, 1918. Mem: ASCE, Am. Railway Engr. Assn. (pres.), Washington Soc. Engrs., Engrs. (pres.), Univ. Club of Chicago, Cosmos Club of Washington DC. Refs: WWE, WWW.

MORSE, EDWARD KIRTLAND; b. Portland, OH, July 3, 1856; d. May 28, 1942; f. Henry Kirtland; m. Mary A. (Lynn); w. Caroline S. Shield; w. 2nd Elizabeth Wood; c. Lucille, Edwina. Grad. from Yale Univ., A.B. 1881. Grad. work at Polytechnic Karlesruhe, Germany. Draftsman Morse Bridge Co. three years, western agent for the co. in Chicago until 1887. Went to Sidney, Australia in 1887 and contracted for building largest bridge in southern hemisphere under firm of Ryland & Morse, Hawkesbury Bridge Co. Gen. contracting work in Pittsburgh, 1889-93. Consulting engr., Jones & Laughlin Steel Co. for twelve years. Consulting engr. in Pittsburgh for four years. Consulting engr. for Allegheny Co., three years. Assn. Flood Commission of Pittsburgh. Transit commissioner to study and report on solutions for transit problem of Pittsburgh. Private practice in Pittsburgh for 20 years. Consulting engr. Western State Hospital for the Insane. Building at Blairsville, PA, 1916-19. Transit commissioner for Pittsburgh, 1916-20. Designed and supt. const. of Carnegie Steel Co. bridges at Port Perry and Rankin across the Allegheny River. Designed and supt. const. of three suspension bridges across Ohio River. Performed many RR and river surveys. Designed and const. mill and mercantile buildings. Chair of City Planning Commission, Pittsburgh. Mem: ASCE (dir.), Engr. Soc. of Western PA (pres.), Am. Inst. of Consulting engrs . Refs: WWE, WWW.

MURPHY, JOHN WILSON; b. New Scotland, NY, Jan. 20, 1828; d. Philadelphia, PA, Sept. 27, 1874; married twice; c. one son and one daughter. Apprentice in mathematics and surveying, received certificate for surveying from William Henry Slingerland, Nov. 15, 1843. Entered RPI, May 5, 1847. Second asst. of engrs. in charge of Western Division of Erie Canal, 1849. Worked under

Squire Whipple, together introduced pin connections in bridges. Designed a suspension bridge with a vertical truss to insure stiffness across the Mohawk River at Tribes Hill, NY. Built levees on the Alabama River, 1851-52. Went into partnership at Trenton, NJ with George Washington Plympton, Assoc. in development of his testing machine, 1855. Began to build bridges on the Murphy-Whipple plan, determined elasticity of the iron as well as its breaking weight, 1856. Bridge building temporarily halted in 1857-59, designed bridges in Philadelphia during this time. Built bridge on Lehigh Valley RR, 1859. Organized a co. with alumni from RPI, const. many important bridges for RR expansion. Chief engr., Montgomery Alabama, 1860-61. Designed pin connected bridge with all wrought iron members for Lehigh Valley RR, 1863. Builder of Union Hall, Philadelphia, June 1864. Built pipe aqueduct over valley of the Wissahickon, 1869. Built Broad St. Bridge in Philadelphia. Refs: RPI Bio, CAB, DAB, WWW.

NETTLETON, EDWIN S.; b. near Medina, OH, Oct. 22, 1831; d. Denver, CO, April 22, 1901; f. Lewis Baldwin; m. Julia (Baldwin); w. Lucy F. Grosvenor; c. four. Attended Oberlin College, 1853-54, left for lack of funds. Went into lumber business at Kalamazoo, MI. Moved to Pleasantville, PA, 1865. Served as county surveyor. Started west, spring 1870. Joined the Union or Greeley Colony at Council Bluffs on its way to CO. Acted as engr. of the colony. Surveyed town site of Greeley and laid out its irrigation ditches. Built the Larimer and Weld Canal for Colorado Mortgage & Investment Co., known as the English Co., largest irrigation system in CO up to that time. Built High Line Canal, known as the English Ditch because it was under financial control of the English Co. Invented weir built for the co. Surveyed town sites of Colorado Springs, 1871. Manitou, 1872. South Pueblo, 1873. Pres. of flour milling co. in Pueblo. State engr. of CO, inaugurated work of gauging streams and ditches, 1883-87. Engaged in consulting work, laid out a number of irrigation works in WY and ID. Chief engr. in diversion of the Yaqui River in Mexico. for irrigation. Appt. consulting engr. for 1st investigation of irrigation of the US govt., 1889-93. Sent to investigate irrigation systems, reforesting denuded tracts, and methods of preventing destruction of forests in Spain in 1889 and 1892, Italy in 1892. Irrigation expert under Dr. Elwood Meade of US Dept. of Agriculture for two years. Made studies in ID, WY and CO, invented instruments for better measurement of water. A founder and 1st trustee of Colorado College at Colorado Springs. Est. weather bureau on Pike's Peak. Became wealthy in real estate boom in Denver, 1898, lost most of his money in the failure of a Denver bank. Refs: DAB, WWW.

NEWTON, JOHN; b. Norfolk, VA, Aug. 24, 1823; d. NYC, May 1, 1895; f. Thomas; m. Margaret (Jordan) Pool; w. Anna M. Starr; c. five sons and one daughter. Attended school in Norfolk, studied civ. engr. under a private tutor. Entered US Military Academy, July 1838, grad. second in his class, July 1, 1842. Commissioned 2nd lt. of engrs. Asst. to bd. of engrs., 1842-43. Asst. prof. 1843-44. Principal asst. prof. of engr. at US Military Academy, 1844-46. Asst. engr. in const. Ft. Warren, MA, 1846, and Ft. Trumbell, CT, 1846-49. Asst. engr. in const. Fts. Porter, Niagara and Ontario, NY, and Ft. Wayne, MI, 1849-52. On surveys for breakwater at Owl's Head in ME, 1852-53. In FL improving St. John's River and at Fts. Pulaski and Jackson, GA, 1853-54. Supt. improving lighthouses on Savannah River. Const. fortifications at Pensacola, FL, 1855-58.

Promoted to 1st lt. on Oct. 16, 1852 and capt., July 1, 1856. Mem. of bd. to examine floating dock at Washington Navy Yard, and of special bd. of engrs. for modifying the plans of ft. at Sandy Hook and for selecting sites for additional batteries at Ft. Hamilton. Chief engr. of UT expedition, 1858. Supt. engr. in const. of Ft. Mifflin, DE, 1858-61. Mem. of special bd. of engrs. on harbor defenses, NY Harbor, 1860. Chief engr. of Dept. of Pennsylvania, 1861. Chief engr. of the Dept. of the Shenandoah, July-Aug. 1861. Promoted to maj., Aug. 6, 1861. Brig. gen. of Volunteers, Sept. 23. Engr. in const. defenses of Washington, DC from Aug. 28, 1861-March 1862. Const. Ft. Lyon. Covered retreat of Pope's Army from Bull Run to Washington, DC, Sept. 1-2, 1862. Served in Maryland Campaign, Sept.-Nov. 1862. Recommended for promotion to maj. gen. for gallantry and services at battle of Antietam, Sept. 17, 1862. Commanded division at Fredericksburg and Chancellorsville from Dec. 13, 1862-June 1863. Commanded his division at Gettysburg, July 1-3, 1863. In operations in GA around Dalton and Adairsville and battles of Dallas, Kennesaw Mountain, Peach Tree Creek, Jonesboro and Lovejoy's Station. Maj. gen., US Army, March 13, 1865. At end of the war, became lt. col. of engrs., Dec. 28, 1865. Mustered out of volunteer service, Jan. 15, 1866. Returned to fortification, river and harbor work. Removed obstructions in East River, NY. Removed a large submerged stone which had caused many wrecks and previously had been unable to be removed. Blew up Flood Rock or Middle Reef in Hell Gate, Oct. 10, 1885. Promoted to col. on June 30, 1879. Became brig. gen. and Chief of Engrs., March 6, 1884. Retained charge of Hell Gate operations until Dec. 31, 1885. Retired at own request, Aug. 27, 1886. Commissioner of Public Works, NYC, Aug. 28, 1886 for two years. Declined re-appointment. Hon. degree of LL.D. of St. Francis Xavier College, 1886. Pres. of Panama RR Co. until his death. Mem: ASCE (hon. mem.), National Academy of Sciences. Refs: DAB, WAB, CAB, Lamb, WWW, DR, BDNA.

NICHOLS, OTHNIEL FOSTER; b. Newport, RI, July 29, 1845; d. Brooklyn, NY, Feb. 4, 1908; f. Thomas Pitman; m. Lydia (Foster); w. Jennie Swazy; c. three. Attended public schools in Brooklyn, NY, prepared for college under Prof. J. C. Fox. Apprenticed as a machinist, 1862-64. Entered RPI, Sept. 1865, grad. with degree of CE, 1868. Asst. in development of Prospect Park, Brooklyn. Employed on 1st elevated RR in NYC. Asst. engr. in office of Cooper & Hewitt, 1870-71. Const. tunnel divisions of Lima & Oroza and of Chimbote & Huaraz RR, Peru, 1871-76. Asst. engr. and supt. for Edge Moor Bridge Works in const. of Metropolitan Elevated RR in NYC, 1876. Engr. in charge of main drainage sewer for annexed district in Park Dept. of NYC. Resident engr. and attorney of Madeira & Manore RR in Brazil and made a trip to London in connection with litigation for the co., 1878-79. Employed by Cooper & Hewitt as asst. engr. in bridge shops of NJ Steel and Iron Co. at Trenton NJ, 1879-81. Asst. to pres. of Peter Cooper Glue Factory in Brooklyn, 1882. Resident engr. of Henderson bridge over the Ohio River, 1882-86. Chief engr. of Westerly, RI water works, 1886. Principal asst. engr. of Suburban Rapid Transit Co. in NY, 1887-88. Chief engr. of Brooklyn Elevated RR, 1888-95, chief engr. and gen. mgr., 1892-95. Principal asst. engr. in charge of Williamsburg Bridge, NYC, 1896-1903. Chief engr., Dept. of Bridges, NY, 1904-1906. Pres. of Dept. of Engr., Brooklyn Inst. of Arts and Sciences. Mem: ASCE (dir., 1892-1893), Am. Assn. of Mech. Engrs., Inst. of Civ. Engrs. of Great Britain, Am. Geographical Soc., Municipal Engrs. of NYC (trustee), Brooklyn

Engrs. Club (pres., 1904). Publ: "The Brooklyn Elevated Railway" (1897). Refs: WWW, RPI Bio, Lamb, WAB, TASCE, BDNA.

NORTHEN, WILLIAM EZRA; b. Amesbury, MA, March 14, 1819; d. April 2, 1897; f. Ezra; m. Mary (Currier); w. Margaret Hobbs. Grad. from Harvard, 1838. Engr. with the Albany & West Stockbridge RR, 1840-42. Formed partnership with George Whistler in an engr. co., 1842-49. Specialized in water problems, dams, floating docks and pumping machinery. Vice pres. and engr. of the NY and New Haven RR, 1854-66. Sanitary engr. with the NY Metropolitan Bd. of Health, 1866-69. Chief engr. of the Chicago Main Drainage Canal, 1890-91. Mem: ASCE (pres., 1897). Publ: *First Lessons in Mechanics* (1862). Refs: WWW.

NOSTRAND, PETER ELBERT; b, Brooklyn, NY, Jan. 15, 1856; f. John Lott; m. Ellen (De Bevoise); w. Ella Frances Arcularius; c. Elizabeth, Elbert Arcularius. Grad. from Brooklyn Polytechnic Inst. with degree of B.S., 1875. Took course in chemistry and assaying in Brooklyn. Principal asst. engr. and dir. with father's co., John Nostrand & Son in design, location and const. of Brooklyn Elevated RR, 1876-80. Superintending engr., Cape Cod Ship Canal, 1880. Mining engr., assayer and asst. engr. to William J. McAlpine, hydraulic work. Chief engr., Ramapo Water Co., 1887-90. City surveyor, NYC. Chief engr., John D. & Thomas E. Crimmens, contractors for const. of 3rd Ave. cable road and Broadway cable road, 1891-93. Consulting engr. for architects on large NY bldgs on foundations including: Columbia Univ. buildings, Cathedral of St. John the Divine, Cherry, Sampson, Hudson, Empire, Woodbridge, Chesebrough, and Mutual Life buildings. Consulting engr. and expert, NYC Ashoken Dam Condemnation. Appt. mgr. of Dept. of Surveys, connected with the Lawyers' Title Insurance Co. of NY, 1902. Consulting hydraulic engr., NY State Barge Canal Condemnation. County engr. and Supt. of Highways, Suffolk County, NY. Dir., Bushwick Savings Bank. Justice of the Peace. Mem: ASCE, Am. Inst. of Mining and Mech. engrs., Brooklyn Inst. for the Arts and Sciences, Univ. Club of Brooklyn. Refs: WWE, WAB, WWW.

OCKERSON, JOHN AUGUSTUS; b. Skane, Sweden, March 4, 1848; d. St. Louis, MO, March 22, 1924; f. Jans; m. Rose Jans (Datler) Atkerson (changed spelling of name when immigrated; w. Helen M. Chapin; w. 2nd Clara Shackelford. Emigrated to the US when two years old. Parents died while making the trip from NY to Chicago. Raised by relatives near Elmwood, IL. Educated in public schools. Enlisted in 132nd Illinois Infantry, 1864. Mustered out after less than six months. Re-enlisted Jan. 1865 in the 1st Minnesota Heavy Artillery, served until the end of the Civil War. In milling business in MN, 1865-68. Entered Univ. of IL, 1869, grad. with degree in civ. engr., 1873. Asst. engr. in location and const. of Atchinson, Topeka & Santa Fe RR, 1872. Principal asst. engr. in the federal Great Lakes Survey, engaged in hydrographic, topographic and triangulation surveys, 1873-79. Surveyed mouth of the Mississippi River const. by James Buchanon Eads, 1876. Prin asst. engr. in charge of surveys and physical examinations from source of river to gulf with Mississippi River Commission, 1879-87. Mgr. and engr., gold and silver mine in CO, 1888-89. Chief asst. engr. in Mississippi River Commission, 1890-1898. Appt. mem. of the Commission Aug. 4, 1898 until his death. Consulting engr. on many projects, including Chicago Drainage Canal, 1907, Big Wood River Dam in

ID, 1909, East Side Levee and Sanitary District, East St. Louis, IL, 1908-1912, Little River Drainage District, MO, 1914, and Conservancy District, Miami Valley, Dayton, OH, 1914. Const. levees to control flood waters of Colorado River, 1910, received personal commendation from Pres. Taft. Degree of D. Engr. from Univ. of Illinois in 1903. Chief Dept. Liberal Arts and mem. of Superior Jury of Awards, St. Louis Exposition, 1902-1905. Delegate from the US to four Intl. Congresses on Navigation, 1900 in Milan, 1905 in St. Petersburg, 1908 in Philadelphia, and 1912 in Paris. Mem. of Intl. Jury of Awards, Paris Exposition. Awards from foreign countries include: Decorated Officer, Public Instruction Meritide Agricole, France, 1900; Knight of the Crown of Italy, 1904; Knight and then Cmdr., Order of Vasa, Sweden; Knight, Crown of Germany; Knight, Crown of Belgium; Order of Double Dragon, China. Mem: ASCE (vice pres., 1907-1908, pres., 1912), St. Louis Engrs. Club (pres.), National Geographic Soc. Clubs. Publ: "Dredges and Dredging of the Mississippi River" (1898), "Stadia Tables, Giving Elevations in Feet," "Out Streams," "The Atchafalaya River," "Levees of the Mississippi River," "River Bank Erosion," "Mechanical Devices for Printing Conventional Signs and Letters on Original Maps." Refs: DAB, WWW, WAB, WWE, TASCE.

JOHN AUGUSTUS OCKERSON (1848-1924)
ASCE PRESIDENT, 1912

OLCOTT, EBEN ERSKINE; b. NYC, March 11, 1854; d. June 5, 1929; f. John Nathaniel Olcott; m. Euphemia Helen (Knox); w. Kate Van Santvoord; c. three sons and one daughter. Educated in public schools, NYC. Attended College of the City of NY. Entered School of Mines of Columbia Univ., grad. 1874. Chemist for Hunt & Douglass Ore Knob Copper Co. in NC, became supt. of plant. Asst. supt. of Pennsylvania Lead Co. at Mansfield Valley, PA, 1876. Supt. of gold mine in Venezuela, 1876-79. Later held similar position in CO.

Supt. of St. Helena Mines, Sonora, Mexico, 1881-85. Opened office as consulting engr. in NY. Explored gold and copper district of eastern Peru, 1890-91. Examined Huantajaya district of Chile, 1892-93. Managed father-in-law's Co. on his death, the Hudson River Day Line, 1895- . Senior member of firm of Olcott, Fearn & Peele, consulting engrs. Trustee, officer and dir. of Catskill Evening Line. Dir., Old Lincoln Bank. Vice pres., Lincoln Safe Deposit Co. Inaugurated movement leading to Tricentennial Hudson-Fulton Celebration in NYC in 1909. Mem: Am. Inst. of Mining Engrs. (pres., 1901-2), Am. Geographical Soc. (council), NY Chamber of Commerce, NY Historical Soc., Merchant's Assn., Engineers Club of NYC, Bd. of Foreign Missions of the Reformed Church in Am., Am. Bible Soc. (mgr.), Am. Seaman's Friend Soc. (trustee), Am. Indian Inst. (treasurer and trustee). Refs: DAB, YPB, WWW, WAB, TASCE, WWE.

OSBORNE, FRANK CHITTENDEN; b. Greenland, MI, Dec. 18, 1857; d. Cleveland, OH, January 31, 1922; f. Reuben Howard; m. Livonia (Chittenden); w. Annie Powell; c. Kenneth Howard. Early education in Calumet High School. Entered RPI, Sept. 1876, grad. with degree in civ. engr. in 1880. Asst. engr. of Louisville Bridge and Iron Co., July 25, 1880, principal asst. engr., Sept. 1880-1885. Principal asst. engr. of Keystone Bridge Co., Pittsburgh PA, 1885-87. Mem. of firm of G. W. G. Ferris & Co., inspectors and designers of structural steel work in Pittsburgh, 1887-89. Chief engr. of King Bridge Co., Cleveland, OH, 1889-92. Opened office as consulting engr., 1892. Incorporated in 1900 as Osborne Engr. Co., pres. from 1900-1910, and 1917-1919, dir. until his death. Built many steel and reinforced concrete bridges for cities and corporations in the US, including the Y bridge over the Licking and Muskingum Rivers at Zanesville, OH, largest reinforced concrete bridge in the US at the time. Vice pres. of American Art Stone Co. Dir. of Lake Shore Banking & Trust Co. Mem. of Cuyahoga County Building Commission, 1908-15. Devised code of conventional signs for bridge riveting. Proposed adoption of standard size sheet of 24 x 36 inches for bridge drawings, became generally used. Mem: ASCE (dir., 1901-1903), Civ. Engrs. Club of Cleveland (dir., secy., 1897, vice pres., pres.,1898), Cleveland Chamber of Commerce, Univ., Athletic and Chippewa (secy. and treas.) Clubs, Inst. of Civ. Engrs. of Great Britain, Western Reserve Historical Soc., New England Soc. Sons of Am. Revolution (pres.), Assn. of Engr. Societies (mgr.), Assn. of Railway Supts. of Bridges and Buildings, Am. Soc. for Testing Materials, Masons, Sigma Xi, New England Soc. of Cleveland, Western Reserve, Western Reserve Historical Soc., Navy League, Aerial League of Am., National Geographic Soc. Publ: Tables of Moments of Inertia and Squares of Radii of Gyration (1894). Refs: WWW, RPI Bio, TASCE, WAB, WWE.

PAGE, LOGAN WALLER; b. Richmond, VA, January 10, 1870; d. Dec. 9, 1981; f. _____ ; m. _____ ; w. Anne P. (Shaler), Oct. 1903; s. Leigh R. and P. (Waller); ed. Powder Point School, Bear Island Academy, grad. from Virginia Polytechnic Inst. 1889, then attended Harvard Univ. and became one of the 1st three Harvard engr. students to specialize in highway engineering. In 1893 he was appointed to direct the Lawrence Scientific School's testing laboratory, whose facilities were available to the MA State Highway Commission. In 1900 he joined the Office of Public Road Inquiry (OPRI; later the Bureau of Public Roads) to direct OPRI's 1st laboratory which operated in conjunction with the Agriculture Department's Bureau of Chemistry. Through his efforts the laboratory very quickly strengthened OPRI's national position as the standard source of information on road materials and const. needs. In 1905 he was elevated to head the reorganized Office of Public Roads, a position which he held until his death. He developed a number of initiatives to place highway design, const. and materials on a more rational and scientific basis. He initiated a series of economic studies to evaluate the importance of good roads, est. demonstruction projects in many states and while the Office of Public Roads served as a national road materials testing facility, Page promoted and assisted states in developing this testing capability for themselves. He and his staff presented many technical papers and participated actively in engineering and professional soc. meetings and had great influence on specifications. In 1904 he became chair of the Am. Soc. for Testing Materials's newly-formed Committee on Road Materials. While attending a meeting of Am. Assn. of State Highway Officials, he suffered a fatal heart attack. Publ: The Testing of Road Materials (1901); Roads, Paths, and Bridges (1912). Refs: HN.

PALFREY, JOHN CARVER; b. Boston, MA, Dec. 25, 1833; d. Boston, MA, Jan. 29, 1906; f. John Gorman; m. Mary Ann (Hammond); c. John Gorham, Francis Winslow, Hannah Gilbert. Attended Boston Latin School. Grad. from Harvard, 1853. Grad. from US Military Academy, 1st in class of 1857. Appt. brevet 2nd lt., then second lt. in the Corps of Engrs. Asst. to bd. of engrs. for Atlantic Seacoast Defenses. Connected with const. and repair of fts. in Portland Harbor ME and Portsmouth, NH. At start of the Civil War, ordered to Ft. Monroe, VA as asst. engr. Supt. engr. in const. of ft. at Ship Island, MS. In charge of const. and repair of fortifications around New Orleans, field works of the Dept. of the Gulf and defenses of Pensacola, FL. Participated in Red River Campaign, 1864. Also participated in operations against Port Hudson, LA, Ft. Gaines, Ft. Morgan and Mobile, AL. Surveyed the Red River, determined practicability of engr. expedients by which the water level was raised, allowing vessels to pass over rapids and escape capture. Immediate charge of field works in operations against Ft. Gaines and Ft. Morgan. Brevetted maj. 1864, lt. col. of Volunteers in 1865, col. and brig. gen. Took part in reconst. of San Antonio and Mexican Gulf RR of TX. Resigned from Army, May 1, 1866. Agent of Merrimack Manufacturing Co. of Lowell, MA. Treasurer of Manchester Mills of Manchester, NH, July 1, 1874-91. Overseer of Thayer School of Civ. Engr. of Dartmouth College. Mem: Phi Beta Kappa Soc. of Harvard, Massachusetts Historical Soc., Military Historical Soc. of Massachusetts (secy.). Publ: "The Siege of Yorktown" (1881), "Port Hudson" (1910). Refs: DAB, WAB.

PARDEE, ARIO; b. Chatham, NY, Nov. 19, 1810; d. Ormond, FL, March 26, 1892; f. Ariovistus; m. Eliza (Platt); w. Elizabeth Jacobs; w. 2nd Anna Maria Robinson; c. fourteen. Moved shortly after birth with his family to a farm in Stephentown, NY. Attended district school, studied under a private tutor. Rodman for surveyor locating Delaware & Raritan Canal in NJ, starting in June 1830. Continued work with the engrs. const. the canal until 1832. Chief engr. to locate Beaver Meadow RR connecting coal mines at Beaver Meadow, PA with the Lehigh Canal at Mauch Chunk, PA, 1832-36. Founded city of Hazelton, PA in 1836, settled there in 1840. Became largest shipper of anthracite coal properties in the state. Chief engr. of Hazleton RR and Coal Co., 1836-40. Independent coal operator in firm of Pardee, Miner & Co. Dir. of Lehigh Valley and other RRs. Built a gravity RR to Penn Haven in 1848, abandoned in 1860 for the Lehigh Valley RR. Owned blast furnaces in Stanhope, NJ, and others in NY, VA and TN. Owned forest land in Canada and NC. Pres. of Secaucus and the Musconetcong Iron Works in NJ, Allentown Rolling Mills and Union Iron Works of Buffalo, NY. Dir. of Lehigh Valley RR and other RRs. Fitted out at his own expense a co. of US Volunteers for the Civil War, 1861. Donated money needed to save Lafayette College in 1864, became trustee in 1865. Endowed the Pardee Scientific Course in 1866. Donated capital needed to build Pardee Hall, 1871. Pres. of the Bd., 1882-92. Presidential elector in 1876. Chair of Bd. of Commissioners for 2nd PA Geological Survey. Refs: DAB, WWW, Lamb, CAB, WAB, BDNA.

PARSONS, WILLIAM BARCLAY; b. NYC, April 15, 1859; d. NYC, May 9, 1932; f. William Barclay; m. Eliza (Livingston); w. Anna DeWitt; c. Sylvia Caroline, William Barclay, Jr. Went to school in Torquay, England, 1871. Studied under private tutors and traveled through France, Germany and Italy, 1871-75. Came back to the US, entered Columbia College, grad. with degree of A.B., 1879, received degree of civ. engr. in 1882. Engr. for Blossburg (PA) Coal Co., summer 1881. Maintenance of Way Dept. of NY, Lake Erie & Western RR, 1882-85. Began practice as consulting engr. with his brother in NY, 1886 for twenty years. Studied plans for an underground RR in the city. Engaged in other RR lines located in the west, Jamaica and British West Indies, and water

WILLIAM BARCLAY PARSONS (1859-1932)
IN 1905 FOUNDED WHAT IS NOW PARSONS, BRINCKERHOFF,
QUADE & DOUGLAS, INC.

supply works including those in Vicksburg and Natchez, MS. Chief engr. of Ft. Worth & Rio Grande RR in TX. Deputy chief engr. of Rapid Transit Commission in NY under William E. Worthen, 1891. Chief engr. under the appt. of new commission with broader powers, 1894-1904. Commission ended activities in 1898 under adverse political pressure. Entered service as capt. in a Volunteer Engr. Regiment. Promoted to Chief Engr., Engrs. of the NY National Guard with rank of brig. gen. Placed in command of regiment at Peekskill, NY to teach principles of army engr. Directed survey of 1000 miles of RR in China on line from Hankow to Canton, 1898-99. Called back to NY by Transit Commission late in 1899. Actual const. of subway const. in NY started in 1900. Completed 1st section in 1904, resigned as chief engr. Assoc. in design of the Spier Falls Plant on the Hudson River, 1900-1902. Appt. to the Isthmian Canal Commission, 1904-1906. Went to Panama early in 1905 as mem. of the Committee of Engrs., reported in favor of a sea level canal. Appt. to the Intl. Bd. of Consulting Engrs. to study types of canal. Appt. to mem. on bd. to consider plans of the Royal Commission on London Traffic, 1904. Consulting engr. for hydroelectric development on Susquehanna River at McCall Ferry, PA, 1905. Consulting engr. to Massachusetts RR Commission. Advisory engr. on traffic problems to Cambridge, San Francisco, Toronto, Detroit and other cities. Consultant on large hydraulic works such as Salmon River, McCall Ferry and Mohawk hydroelectric developments. Organized firm of William Barclay Parsons, consulting engrs. in partnership with Eugene Clapp, 1905. Const. of Steinway Tunnel under East River in NY, 1905. Firm supervised const. of hydroelectric plant at Colliers, NY on the same river, 1906-1907, designed hydroelectric plant at Ephratah, NY on Garoga Creek, 1908-11, acted as consulting engrs. to the Niagara, Lockport and Ontario Power Co., 1911-14, made report and plans for the reclaimation of Zapata Swamp in Cuba. Chief engr. of Cape Cod Canal, 1st canal without locks between waters having considerable tidal differences, 1905-14. Firm changed name to Barclay Parsons and Klapp, 1909. Changed again in 1920 to Parsons, Klapp, Brinckerhoff & Douglas. John P. Horgan admitted to partnership in 1926. LL.D. from St. Johns College in MD, 1909. Studied and reported on urban and interurban transit for San Francisco, CA, Detroit, MI, Cambridge, MA, Baltimore, MD, Chicago, IL, Philadelphia, PA, Toronto, Ontario, Canada, and other cities, 1904-17. Chair of Chicago Transit Commission, 1916-17. Chair of Joint Commission of Engr. Societies appt. to report on military engr. problems and requirements for engr. troops, 1915. Served with 11th US Engrs. in England as maj., 1916, lt. col., 1917, and col., 1918 until end of the war. Regiment const. four-track RR that carried Am. troops from French coast to the front. Participated in engagement at Cambrai, the Lys defensive, the Saint Mihiel and Argonne-Meuse campaigns. Received Distinguished Service Medal, Victory Medal and Citation for Conspicuous Service from the US, Distinguished Service Order from Great Britain, Legion of Honor from France, Order of the Crown from Belgium. Transferred to the Engrs. Reserve Corps after the war with rank of brig. gen. Consulting engr. for NY Water Power Investigation, 1920-23. Built intl. vehicular tunnel passing under the Detroit River which joined Detroit with Windsor, Ontario, Canada which opened in 1930. Trustee of Carnegie Inst. Mem. of Bd. of Trustees for Columbia Univ., 1897, chair in 1917. Trustee of NY Public Library. Trustee of Carnegie Inst. of Washington, DC. Chair, Administrative Bd. of Columbia-Presbyterian Medical Center in NY. Mem: ASCE (hon. mem., dir., 1896-98), Inst. of Civ. Engrs. of Great Britain (Telford

Gold Medal), Am. Academy of Arts and Sciences, Societe des Ingenieurs Civils of France, Inst. of Consulting Engineers, Am. Inst. of Architects (hon. mem.). Publ: *Track; A Complete Manual of Maintenance of Way* (1885), *Turnouts; Exact Formulae for their Determination* (1885), *An American Engineer in China* (1900), *The American Engineers in France* (1920), *Robert Fulton and the Submarine* (1922), article in *Encyclopedia Britannica* for 11th edition on "Rapid Transit". Refs: DAB, Lamb, WWE, BDNA, WWW, TASCE, YPB.

PETERS, RICHARD; b. Germantown, PA, Nov. 10, 1810; d. Atlanta, GA, Feb. 6, 1889; f. Ralph; m. Catherine (Conyngham); w. Mary Jane Thompson; c. nine. Moved with family to Wilkes-Barre, 1821, and to Bradford County, 1823 or 24. Worked on a farm. Went to Philadelphia, studied math, drawing, and writing for 18 months. Entered office of William Strickland for six months. Asst. in const. of Delaware Breakwater. Asst. engr. in location of the Camden & Amboy RR. Asst. engr. of the Georgia RR, 1835-37. Worked on Charlestown & Hamburg RR. Supt. of Georgia RR, 1837-46. Resigned in 1846 on completion of Georgia RR to Marthasville which name he changed to Atlanta. Bought stage line from Madison, GA to Montgomery, AL, 1844-50. Bought large plantation in Gordon County, GA, 1847. Built largest flour mill in GA, 1856. Pres. of GA Western RR, 1860-65. Dir. of Atlanta & West Point RR, 1865-89. Lessee and dir. of Western & Atlantic RR 1870-89. Est. St. RR in Atlanta, GA, 1871. Mem. of Atlanta City Council. Commissioner, Gordon County. Refs: WWW, DAB, WAB.

PLYMPTON, GEORGE WASHINGTON; b. Waltham, MA, Nov. 18, 1827; d. Sept. 11, 1907; f. Thomas Ruggles; m. Elizabeth (Holden); w. Delia M. Bussy; w. 2nd Helen M. Bussy; c. Harry and three daughters. Early education at district school and at high school in Waltham. Worked in a machine shop, 1844-47. Prof. of chemistry and toxicology at Long Island College Hospital, 1844-45. Grad. from RPI with degree of civ. engr., 1847. Inst. of Geodosy and Mathematics at Renssealaer Polytechnic Inst., 1847-48. Civ. engr. work in MA, NY, OH, 1848-52. Prof. of engr. and architecture in Western Reserve Univ., Cleveland, OH, 1852-53. Prof. of mathematics at NY State Normal College, 1853-55. Hon. A.M. from Hamilton College, 1854. Opened firm of Plympton and Murphy, building bridges in Trenton, NJ, 1855-57. Prof. of physics and engr. at the State Normal School at Trenton, NJ, 1857-9. Engaged in engr. work for US govt. during Civil War. Chair of physics and engr. at Brooklyn Polytechnic Inst., Brooklyn, NY, 1863-79. Hon. degree of M.D. from Long Island College Hospital, 1877. Dir. of Night School at Cooper Union, NYC, 1879. Prof. of chemistry and toxicology at Long Island College Hospital, 1864-86. Commissioner of Electrical Subways of Brooklyn, 1885-89 and 1892-96. On bd. of experts selected in 1890 to improve method of transportation across Brooklyn Bridge. Planned water supply and sewerage systems for Bergen, NJ. Surveyed the marl deposits of NJ and the const. of iron bridges. Editor of *Van Nostrand's Engineering Magazine*, 1870-86. Mem: ASCE. Publ: *The Blowpipe, a Guide to its Use in the Determination of Salts and Minerals* (1858), *A Translation of Jannettaz's "Guide to the Determination of Rocks"* (1877), *The Starfinder or Planisphere with a Movable Horizon* (1878), *The Aerinoid and How to Use It* (1880), *How to Become an Engineer* (1892), *Surveying*, "Some Experiments on the Transverse Breaking Strain of Plate Glass." Refs: TASCE, Lamb, WAB, BDNA, CAB, RPI Bio, WWW.

POE, ORLANDO METCALFE; b. Navarre, OH, March 7, 1832; d. Detroit, MI, Oct. 2, 1895; f. Charles; m. Susanna (Warner); w. Eleanor Carrol Brent; c. four. Entered the US Military Academy, 1852. Grad. 6th in class of 1856. Appt. brevet 2nd lt., Corps of Topographical Engrs., Oct. 7, 1856. Asst. topographical engr. on survey of northern lakes, 1856-61. 1st lt., July 1,1860. Asst. in organizing Ohio Volunteers, 1861. Chief topographical engr., Dept. of the Ohio, May 13-July 15, 1861. Mem. of staff of Gen. McClellan at Washington, DC, July-Sept. 1861. Appt. col. of 2nd Michigan Volunteers, Sept. 16, 1861. Volunteer commission expired March 4, 1863. Chief engr. of Central District of KY, April-June 1863. Chief engr. of XXIII Army Corps march on Knoxville, TN, June-Sept. 1863. Chief engr. of Army of the Ohio, Sept.-Dec. 1863. Planned and const. fortifications of Knoxville. Brevetted rank of maj. Assigned to asst. engr. of Military District of the Mississippi, Dec. 1863-April 1864. Selected as chief engr. of Sherman's Army, April 1864. Brevetted lt. col., Sept. 1, 1864, and col., Dec. 21, 1864. After the war, served as engr. secy. of the Lighthouse Bd., 1865-70. Commissioned maj. in Corps of Engrs., March 7, 1867. Engr. of Upper Lakes Lighthouse District and supt. of river and harbor works in Lake region. Built Spectacle Reef Light, Lake Huron, 1870-73. Appt. col. and aide de camp to Gen. Sherman 1873-84. Promoted lt. col., Corps of Engrs., 1882. Col., 1888. Supt. engr. of improvement of rivers and harbors on Lakes Superior and Huron, and of St. Mary's Falls Canal, 1883. In charge of improvement of St. Mary's and Detroit Rivers, ship channel between Chicago, Duluth, and Buffalo, const. of the dry dock, St. Mary's Falls Canal, and design and const. of the locks at Sault Ste. Marie. Publ: "Personal Recollections of the Occupation of East Tennessee and the Defenses of Knoxville " (1889), *Ordnance Notes--No. 345. Report on Transcontinental Railways* (1883). Refs: DAB, WWW, Lamb, BDNA, WAB, CAB.

PRATT, THOMAS WILLIS; b. Boston, MA, July 4, 1812; d. Boston, MA, July 10, 1875; f. Caleb; m. Sally (Willes or Willis); w. Sarah Bradford; c. one son and one daughter. Early education in Boston Public Schools. Attended Rensselaer Academy at Troy, NY (later RPI). Asst. father in work as architect and builder. Engr. asst. with US govt. on const. of dry docks at Charlestown, SC and Norfolk, VA, 1830. Engaged with Boston & Lowell and Boston & Maine RRs. Division engr. on const. of Norwich & Worcester, subsequently supt. of the road, 1835. Engr. and supt. of Providence & Worcester, 1845-47, and of Hartford & New Haven, 1847-50. Chief engr. of Middletown Branch RR. Chief engr. and supt. of NY & Boston RR. Chief engr. and supt. of Conway & Great Falls branch of Eastern (Boston & Maine) RR, 1871-75. Built many important bridges, largest over Merrimac at Newburyport, MA, 1865. Invented Pratt bridge and roof truss, patented April 4, 1844. Patented a new type of steam boiler, Sept. 26, 1865; an equalizer for drawbridge supports, Feb. 22, 1870; an improved type of combined timber and steel truss, April 1, 1873; a new method of hull const. for ships, May 4, 1875 and a new method of propulsion, June 1, 1875. Refs: DAB, WWW, WAB.

PURDON, CHARLES DE LA CHEROIS; b. Belfast, Ireland, Oct. 6, 1850; f. Charles de la Cherois; m. Jane Maria (Calvert); w. Jennie Theo Arthur; c. Arthur, Eleanor D. Educated in private schools in Belfast and Queen's Univ. Studied univ. course under a private tutor. Axman, then asst. engr. for Intercolonial RR in Canada, 1876. Asst. engr. in Public Works Wept. of Canadian govt. on St. Lawrence River, 1872-75. Surveyor in TX, 1876-80. Asst. engr. on Texas & St. Louis RR, 1880-84. Resident engr. for Little Rock Junction RR, 1884-85. Asst. engr. for St. Louis and Santa Fe RR, 1886-87. Asst. chief engr. of St. Louis, Arkansas & Texas RR, 1887-88. Resident and division engr. of L. & N. RR, 1888-90. Asst. on canal bridge, Duluth, MN, June-Aug, 1890. Bridge engr., Atchinson, Topeka and Santa Fe RR, 1890-93, resident engr. of Chicago Division, 1893-95, principal asst. engr., 1895-97, asst. chief engr., 1897-1901. Chief engr. of Kansas City Belt RR, Jan-May 1901. Chief engr. of St. Louis & Santa Fe RR, 1901-1904. Engr., maintenance of way, 1904-1906. Consulting engr., 1906-1909. Chief engr. of Memphis Rd. Terminal, April-Sept., 1909. Asst. engr. of Missouri P RR on valuation of Nebraska lines and grade separation in St. Louis, 1909-10. Chief engr. of the St. Louis Southwestern RR, 1910-18. Consulting engr. for the St. Louis Southwestern RR. Mem: ASCE, Am. Railway Assn., Am. Soc. for Testing Materials, Mason, Engrs. Club, Railway Club. Publ: articles to technical journals. Refs: WWW, WWE.

PURDY, CORYDON TYLER; b. Grand Rapids, WI, May 17, 1859; d. Melbourne, FL, Dec. 26, 1944; f. Samuel Jones; m. Emma Jane (Tyler); w. Eugenia Cushing; w. 2nd Rose Evelyn Morse; c. Corydon Phillips. Educated in public schools, taught in a village school for two years. Entered Univ. of Wisconsin, 1881. Left after one year. Draftsman of const., then asst. engr. of new line of the Chicago, Milwaukee & St. Paul RR between Chicago and Evanston, 1880-83. Returned to Univ. in 1883, grad. in 1885 with degree of civ. engr. One year of graduate study in civ. engr., 1886. City engr. of Eau Claire, WI, 1886-87. Engr. for Keystone Bridge Co., 1888-89. Partnership in Chicago with Charles G. Wade, 1889. Built structural system of Rand McNally Building in Chicago, 1st all steel skeleton ever used. Partnership with J. N. Phillips, 1891-92. Founded firm with Lighter Henderson, name of Purdy and Henderson, 1893-1916, NY office opened, 1894, est. business in Cuba, 1900, incorporated 1901. Est. business in Canada, 1909. Designed iron frames of the twelve story Boyce Building and fourteen story Ellsworth Building, completed in 1892. Built skeleton of Old Colony Building, 1894. Built Marquette building in Chicago, 1894. Opened NY office in 1894, branch offices in Boston, Montreal, Seattle, and Havana. Built NY skyscrapers, Fuller Building, 1903, and Metropolitan Tower, 1909, tallest building in the world when it was finished. Managed office of George Fuller, a building contractor in Chicago, 1896-97. Prepared an exhibit on steel frame const. for the Paris Exposition, 1900. Consulting engr. for revision of Baltimore's building code, 1904. Retired as pres. of his firm in 1917. Chair of Bd. of Dir. until his death. Mem. of Bd. of Dir., National Child Welfare Association. Mem: ASCE, Western Soc. of Engrs., Inst. of Civ. Engrs. of Great Britain, Engineers Club of NY, Commonwealth of Montclair, NJ (hon. mem.), Arctic Club of Seattle, WA (life mem.). Publ: "The Times Building" (Telford Premium, 1909). Refs: DAB, WWW, WAB, WWE.

RAFTER, GEORGE W.; b. Orleans, NY, Dec. 9, 1851; d. on visit to Karlsbad, Austria, Dec. 29, 1907; f. John; m. Eleanor (Willson); w. Alyda Kirk; c. Ethel, Myra Willson. Educated in Phelps Union School, NY and in old Canandaigua Academy. Enrolled at Cornell Univ. for a short time. Principal of Phelps Union School, 1872-73. Asst. part time in City Surveyor's Office, Rochester, NY, 1874-5. Studied architecture in office of A. J. Warner. Taught mathematics in a private school conducted by Prof. Reed. Asst. engr. of Rochester Water Works, 1876. Consulting civil engr. for some cos., 1877-78, including Rochester & Lake Ontario RR Co. and development of coal property in McKean county, PA. Engr. on const. of Texas & Pacific and Missouri Pacific RRs, 1880-82. In charge of water supply of the Texas Pacific RR across the "Staked Plains," 1881. Const. water works at Ft. Worth, TX, 1882-83. Asst. engr. of Rochester Water Works, 1883-87. Engr. of Fredonia Water Works, 1883-90. Surveyed Honeoye Lake for use as storage reservoir for water power for Rochester Mills, 1883. In charge of additional water supply for Rochester, NY, June 1888-Oct. 1890. Expert in sanitary engr. on Boston Water Works, 1889-90. Chief engr. of Rochester Water Works, 1890. Designed sewage disposal plants at Albion, Holley, NY and the WV State Hospital for the Insane, 1890-91, also designed plants at Lawrenceville and Geneva, NY. Designing and const. engr. for water works at Berwick and Nescopeck, PA. Served as consulting engr. for Warsaw NY Water Co. Collaborated with Prof. William T. Sedgewick in developing Sedgewick-Rafter method of water analysis. One of 1st engrs. to use microscope in the studying the biology of water supplies. Studied problems of river control of NY state, 1893. As engr. in charge, devised system of storage reservoirs to regulate flow of the Genesee River, 1893-97, and the Hudson River, notably the reservoir at Schroon Lake, 1895-97. Supt. const. of dam which formed Indian Lake in a tributary of the upper Hudson. Studied movable bridges and investigated high masonry dams abroad in England, Holland, Germany, Italy and France, 1894. Designed water works for Traverse City, MI, 1897, and high masonry dam on the Indian River, NY for the Indian River Co. Built Indian River Dam, 1898. In charge of water supply investigation of US Bd. of Engrs. on Deep Waterways, 1898-99. Consulting engr. to Committee on Canals of NY, 1899. Made surveys and estimates for ship canal from Lake Ontario through Oneida Lake down the Mohawk to the Hudson River. Made plan for great reservoirs and a 90-mile water supply canal or an alternate 28-mile tunnel. Pioneer study of application of hydraulic formulas to heavy flows over weirs. Began private consulting practice after 1900. Mem. of newly founded Water Storage Commission, 1902. Prepared report on hydrology of NY state for the state geologist, 1904. Mem: ASCE, Yonnondio Masonic Lodge (life mem.), Rochester Engr. Soc., Historical Soc., Chamber of Commerce, Oak Hill Country Club, Masonic Club, Cornell Club of Rochester, Am. Forestry Assn., Am. Geographical Soc., Soc. for Protection of the Adirondacks, Soc. of the Genesee, and Am. Water Works Assn. Publ: *Sewage Disposal in the United States* (1894), *Water Resources of the State of New York* (1899), *Hydrology of the State of New York* (1905), and over 175 books and papers and many other reports. Refs: DAB, TASCE, WAB.

RANDOLPH, ISHAM; b. New Market, VA, March 25, 1848; d. Chicago, IL, Aug. 2, 1920; f. Robert Carter; m. Lucy Nelson (Wellford); w. Mary Henry; c. Robert Isham, Oscar De Wolfe, Spottiswode Wellford, George Taylor Randolph. Taught at home and in private schools. Axeman and rodman of Winchester & Strausberg RR, part of Baltimore & Ohio system. Levelman on surveys for Washington and Ohio RR, 1870. Transitman on surveys for extension of Lehigh Valley RR from Jugtown Mountain to Perth Amboy, NJ, 1871. Transitman on extension of Baltimore & Ohio RR to Chicago. March 1872. Resident engr. of the Baltimore & Ohio RR, built 27 miles of track and the roundhouse and shops at South Chicago, IL, 1873. Asst. engr. of Scioto Valley RR. Chief engr. of Chicago & Western Indiana Belt RR, built terminals and freight houses, 1880. Opened an office in Chicago. Engaged in general engr. practice, 1885-93. During this time, chief engr. of locating and building the Chicago, Madison & Northern RR and the Freeport & Dolgeville line. Consulting engr. for Union Stock Yards and Baltimore & Ohio RR. Appt. chief engr. of Sanitary District of Chicago, June 1893-1907. Const. Chicago Drainage canal, largest artificial canal in the world at that time, 1907-1912. Awarded gold medal by the Paris Exposition in recognition of this work, 1900. Acted as consulting engr. for five years after the project's completion. Mem. of bd. of consulting engrs. to determine type of Panama Canal, 1905. Recommended the lock type as opposed to the sea level. Mem. of advisory bd. of six engrs. accompanying Pres.-elect Taft to Panama, 1908. Bd. changed type of canal to lock type. Designed and const. Obelisk Dam above Horseshoe Falls for Queen Victoria Niagara Falls Park Commission. Consulting engr. for Toronto in connection with track elevation and const. of new water supply system and works. Chair of Intl. Improvement Commission of IL. Asst. in plans for canal from Lockport to Utica, IL. Mem. of IL State Conservation Commission, of state River and Lakes Commission and of Chicago Harbor Commission. Engaged in land reclamation work. Served as consulting engr. for Little River Drainage District of southeast MO. Chair of FL Everglades Commission to consider drainage of Everglades. Designed and began const. of harbor system for Miami, FL, 1915. Reviewed plans and estimates for barge canal for Lake Erie and Ohio River Canal, 1916. Hon. degree of D.Engr., Univ. of Illinois, June 15, 1910. D.Com.Sc., Washington and Lee Univ., 1917. Awarded Elliot Cresson Medal from Franklin Inst. of Philadelphia, Feb. 1913. Elected pres. of Citizens Unit of 108th Volunteers during World War I. Mem: ASCE (dir., 1916), Western Soc. of Engrs. (pres.), Am. Inst. of Consulting Engrs., Royal Soc. of Arts of Great Britain. Refs: DAB, TASCE, WWW, WAB.

RANNEY, HENRY JOSEPH; b. Middletown, CT, 1808; d. Lewisburg, LA, May 1, 1865; f. Moses; m. Elizabeth (Gilchrist); unmarried. Attended public schools, grad. from Norwich Univ. in 1828. Chief of party in charge of 3rd Division of Baltimore & Ohio RR, 1831. Appt. principal asst. engr., 1831-32. Principal asst. engr. then chief engr. of the Lexington & Ohio RR in KY, 1832-35. appt. chief engr. of New Orleans & Nashville RR, 1835-42. An early advocate of broad gauge track. Went to Europe to purchase iron and supplies for the road, inspected RRs and rolling stock, 1835. Assoc. with Col. William S. Campbell in engr. work in New Orleans and other places in the south. Chief engr. and pres. of New Canal & Shell Rd Co., New Orleans after 1850. Active in local politics. Promoter of New Orleans, Jackson & Great Northern RR, 1858-62, chief engr. until 1860 then president until his death. Moved to Corinth, MS with the co., Aug. 27, 1862. Served as maj. in the LA Militia for a few years. Refs: WWW, WAB.

RANSOME, ERNEST L.; b. England, _____ 1844; d. _____, 1917; f. Frederick; m. _____; w. _____; c. _____. His great grandfather had est. an iron foundry near Ipswich which remained the family business. His father is credited with the patent that enabled de Navarro and others to perfect the rotary kiln for the manufacture of cement. Ernest moved to US in 1870 and settled with his family in San Francisco. Growing from his assoc. with Pacific Stone Co., he pioneered in a variety of technological innovations in reinforcing, mixing equipment, and construction systems. These he describes in his "Personal Reminiscence." Through his firms, Ransome Concrete Co., Ransome and Smith Co., Ransome Concrete Machinery Co., and Ransome Construction Co., his influence and handiwork were distributed across the country. Publ: *Reinforced Concrete Buildings* (with Alexis Saubrey, 1912). Refs: A. L. Huxtable, "Reinforced Concrete Construction: The Work of Ernest L. Ransome, Engineer 1844-1917," *Progressive Architecture*, Sept. 1957.

RENWICK, HENRY BREVOORT; b. NYC, Sept. 4, 1817; d. NYC, Jan. 27, 1895; f. James; m. Margaret Anne (Brevoort); w. Margaret Janney; c. two. Grad. from Columbia College in 1833. Dry goods clerk in NY for two years. Studied engr., 1835-37. Asst. engr. for US govt., 1837-40. Built breakwaters at Sandy Hook and Egg Harbor, NJ. 1st asst. astronomer with US Boundary Commission, 1840-47. Examiner in US Patent Office in charge of divisions of metallurgy, steam engines, navigation, civ. engr. and ordnance, 1847-52. 1st US Inspector of Steam Vessels at Port of NY, 1853-70. Employed by prominent patent lawyers, took part in almost all the great patent litigations including those of the sewing machine suits, McCormick Reaper and Bell Telephone cases, 1870-95. Publ: *The Lives of John Jay and Alexander Hamilton* (1840). Refs: DAB, BDNA, CAB, WWW, Lamb, WAB.

RICE, GEORGE STAPLES; b. Boston, MA, Feb. 28, 1849; d. NYC, Dec. 7, 1920; f. Reuben; m. Harriet Tyler (Kettell); w. Rose Breuchaud Porter; c. Albert Fteley Rice. Educated in the public schools of Boston. Entered Harvard Univ., grad. 1870 with degree of S.B. Worked during summer of 1869 at the engr. dept. of Boston Water Works, asst. in const. of Chestnut Hill Reservoir. After grad, asst. engr. of the Lowell, MA Water Works, 1870-72. Division engr. of Additional Water Supply, Boston, 1872-77. Asst. engr. and prin asst. engr. in charge of Boston Main Drainage Works, 1877-80. Engaged in mining operations in AZ and CO, 1880-87. Principal asst. engr. of main drainage works of Boston, 1887. Asst. deputy chief engr. of Aqueduct Commission of NYC, 1887-91. Chief engr. of Boston Transit Commission, 1891. Engaged in private practice of engr. with George Evans, under name of Rice and Evans, 1891-1900. Designed and const. water works of New Bedford, MA. Inst. in sanitary engineering at Harvard Univ., 1892-1900. Deputy chief engr. of Bd. of Rapid Transit RR Commissioners of NYC, served William Barclay Parsons in const. of 1st subway of NYC. Succeeded Parsons on his resignation on Dec. 31, 1904 as acting chief then chief engr. of the commission. Engr. of subway const., 1907-10. Private practice, 1910-14. Division engr. in charge of const. for large part of Dual Subway System, 1914-Dec. 7, 1920. Mem: ASCE, Boston Soc. of Civ. Engrs., Am. Inst. of Mining and Metallurgical Engrs., New England Water Works Assn.,

Soc. of Colonial Wars, Univ., Harvard and Arkwright Clubs of NYC, and Union and St. Botolph Clubs of Boston. Refs: TASCE, WWW, WAB.

RICKETTS, PALMER CHAMBERLAIN; b. Elkton, MD, Jan. 17, 1856; d. Baltimore, MD, Dec. 10, 1934; f. Palmer C.; m. Eliza (Getty); w. Viera Conine Renshaw. Moved with his mother after his father's death to NJ, 1868. Educated in private schools and under a private tutor in Princeton, NJ. Entered RPI in 1871, grad. with degree of civ. engr., 1875. Instructor at RPI, 1875-92, asst. prof, 1882-84, prof. of technical mechanics, 1884, dir., 1892-1901, pres. and dir., Feb. 1901-34. Asst. engr. on Troy & Boston RR, summers of 1876-77. Consulting engr. in charge of design and const. of bridges and hydraulic work, 1880-92. Bridge engr. of Troy & Boston RR, 1886-87. Bridge engr., Rome, Watertown & Ogdensburg RR Co., 1887-91. Engr. to public improvement commission of the city of Troy, 1891-92. Engr. of Public Improvement Commission, Corning, NY, 1897-98. Expert in patent cases, 1886-97. Dir. of National City Bank. Trustee and vice pres., Troy Public Library. Trustee of Dudley Observatory, Albany, Albany Academy, Albany Medical College, NY State College for Teachers. Dir. of United National Bank of Troy. Pres. of Lakeview Samaritan Hospital in Troy, 1914-16. Cmdr. of Order of the Crown of Italy, Cmdr. of Legion of Honor, France. Hon. degree of Engr.D. from Stevens Inst. of Technology, 1905. Degree of LL.D. from New York Univ., 1911. Mem: ASCE (dir., 1899-1901, vice pres., 1916-17), Am. Assn. for Advancement of Science, Am. Soc. of Mech. Engrs., Am. Inst. of Mining and Metallurgical Engrs., Am. Philosophical Soc., Inst. of Civ. Engrs. of Great Britain, Troy Club and Union Club of NYC. Publ: *History of the Rensselaer Polytechnic Institute* (1914, 2nd ed.). Refs: WWW, WAB, WWE, RPI Bio, DAB.

ROBERTS, BENJAMIN STONE; b. Manchester, VT, Nov. 18, 1810; d. Washington, DC, Jan. 29, 1875; f. Martin; m. Betsey (Stone); w. Elizabeth Sperry; c. three. Educated in the common schools. Grad. from the US Military Academy in 1835. Brevetted 2nd lt. in the 1st Dragoons, July 1, 1835. Frontier service in IA and KS, 1835-38. Commissioned 2nd lt., 1st Dragoons, May 31, 1836. 1st lt., 1st Dragoons, July 31, 1837. On recruiting duty, 1838-39. Resigned Jan. 28, 1839. Chief engr. of Champlain & Ogdensberg RR, Jan. 1839-40. Asst. geologist of NY state, 1841. Asst. to George Washington Whistler in const. RR from St. Petersburg to Moscow, Russia, 1842. Returned to US and studied law. Practiced in Des Moines, IA, 1843. Lt. col. of state militia, 1844-46. Reappointed 1st lt with Mounted Rifles in Mexican War, May 27, 1846. Promoted capt., Feb. 16, 1847. Participated in battles of Gen. Scott's campaign. Brevetted maj. on Sept. 13, 1847 for gallantry in taking of Chapultepec. Brevetted lt. col. for gallantry Nov. 24, 1847, in action against Mexican guerillas near Matamoras. Returned to the US, on frontier duty in KS and Dakota Territory, 1848-49. Received a sword of honor from the state of IA, Jan. 15, 1849. On leave of absence, 1850-52. Examined land titles in Topographical Bureau at Washington, DC, 1852-53. On leave of absence and on frontier duty in TX and NM, 1853-61. Promoted to maj., 3rd Cavalry on May 13, 1861. Commanded Southern Military District of NM under Gen. Canby, 1861-62. Participated in engagements at Ft. Craig, Albuquerque, Valverde and Paralta. Brevetted col. on Feb. 21, 1862 for gallant and meritorious services at Valverde. Promoted brig. gen. of Volunteers July 16, 1862. Mem. of Gen. Pope's staff. Engaged in battles of Cedar Mountain, Rappahannock Station, Sulphur Springs, and Second Manassas. Brevetted brig. gen. and maj. gen. for gallantry in this campaign, March 13, 1865. Led expedition against hostile Chippewa Indians in Minnesota, Nov. 1862. Commanded unit in the defenses of Washington, DC and an independent brigade in western VA and the District of IA, 1863. In command of 1st Division, XIX Corps in LA, summer 1864. Chief of Calvary in Dept. of the Gulf, Oct. 1864-Jan. 1865. Commanded District and Cavalry Division of West TN, 1865. Discharged from Volunteer Service, Jan. 15, 1866. Brevetted brig. gen. of regular Army for services at Cedar Mountain and maj. gen. of Volunteers for that action and the second battle of Bull Run, March 13, 1865. Promoted to lt. col., 3rd Cavalry, July 28, 1866. Served frontier duty in NM, 1867-68. Instructor in military science at Yale College, 1868-70. Retired from active service, Dec. 16, 1870. Practiced law and prosecuted claims for the government in Washington, DC. Organized stock co. to finance manufacture of a breech-loading rifle which he invented. Co. was unsuccessful. Publ: *Description of Newly Patented Solid Shot and Shells for Use in Rifled Ordnance* (1864), *Lieutenant-General U.S. Grant* (1869). Refs: DAB, CAB, Lamb, BDNA, WAB, DR.

FIVE PAIRS OF LOCKS ON THE ERIE CANAL AT LOCKPORT, NY PLANNED AND CONSTRUCTED BY NATHAN S. ROBERTS

ROBERTS, NATHAN SMITH; b. Piles Grove, NJ, July 28, 1776; d. near Lenox, NJ, Nov. 24, 1852; f. Abraham Roberts; w. Lavinia White; c. at least one son, Nathan Smith. Purchased 100 acres of land in VT and cut timber. Returned to NJ and taught school in Plainfield during the winter. Moved to Oneida County, NY, 1804. Taught school at Oriskany 1804-1906. Appt. principal of the academy at Whitesboro, 1806-16. Purchased farm in Lenox, 1816. Asst. to Benjamin Wright of Rome, engr. in charge of building the middle section of the Erie Canal, 1816-22. Survey of the route, July 1816. Exploratory surveys and location work on the section of canal between Rome and Rochester. In charge of const. of the western section of canal from Lockport to Buffalo, 1822-25. Drafted plan which was adopted and supt. const. of five pairs of locks at Lockport to overcome barrier of a 60-ft. rocky ridge. Consulting engr. for Chesapeake & Delaware Canal, 1825-26. Made a survey and reported on route for a ship canal around Niagara Falls, 1826. Chief engr. of western end of Pennsylvania Canal between Pittsburgh and Kiskimenetas, 1826-28. Investigated and reported on the practicability of supplying the summit level of the projected Chenanago Canal with water for the NY State Canal Bd. Reviewed estimates of line of the Chesapeake & Ohio Canal for the Secy. of War. Examined country between Johnstown and Franktown for a possible route for a RR or portage over the mountains to connect eastern and western sections of the Pennsylvania State Canal. Mem. of bd. of engrs. of Chesapeake & Ohio Canal Co., Dec. 1828-30. Revised and located expansions of that canal. Built section between Point of Rocks and Harpers Ferry. Assoc. with Jonathan Knight in preliminary work in same territory for Baltimore & Ohio RR. Appt. by the federal govt. to take charge of surveys for a ship canal around Muscle Shoals in Tennessee River, AL. Assoc. with John B. Jervis and Holmes Hutchinson for series of examinations and surveys for the NY State Canal Bd., estimated expense of enlarging the Erie Canal, spring 1835. Chief engr. of enlarging the Canal between Rochester and Buffalo, 1839-41. Removed from office by the Whig administration, 1841. Retired to his farm. Refs: DAB, WWW, GEPT.

ROBINSON, ALBERT ALONZO; b. South Reading, VT, Oct. 21, 1844; d. Topeka, KS, Oct. 7, 1918; f. Ebenezer, Jr; m. Adeline (Williams); w. Julia Caroline Burdick; w. 2nd Ellen Frances (Burdick) Williams; c. Metta Burdick. Mother remarried in 1853 after his father died in 1848. Moved to the west with his family. Educated in public schools and grad. from academy at Milton, WI. Entered Univ. of Michigan, grad. in 1869 with degrees of B.S. and civ. engr., M.S. in 1871. Worked as a clerk in stores, 1856-59. Asst. engr. with US Lake Survey for five months every year while a student at univ., 1866-68. Joined engr. corps of St. Joseph & Denver City RR, May 27, 1869. Axeman and asst. engr. until April 1, 1871. Asst. engr. in charge of location and const. with Atchinson, Topeka & Santa Fe, April 1, 1871; chief engr., April 1, 1873-Aug. 1890; division supt., 1880-81; supt. of bridges, buildings and water service, Oct. 1, 1881-June 1, 1883; asst. gen. supt., June 1-Sept. 1, 1883; gen. supt., Sept. 1, 1883-March 1, 1884; gen. mgr., March 1, 1884-Feb. 1, 1886; 2nd vice pres., Feb. 1, 1886-May 1888, 2nd vice pres. and gen. mgr., May 1888-May 3, 1893. In charge of occupation of Raton Pass a few hours ahead of the force sent out by rival Denver & Rio Grande, early in 1878. Secured for Santa Fe the only practical gateway into northern NM. Designed and const. the switchback which the 1st Santa Fe locomotive used to cross the summit of the Pass and enter the region before the completion of the tunnel. Pres. of Mexico Central RR Co., Ltd, 1893-

Dec. 1, 1906. Mem: ASCE, Topeka Chamber of Commerce (pres.), Topeka Club, Country Club, Chicago Club, Lawyers' Club of NYC, and Jockey and Univ. Clubs of Mexico City. Refs: DAB, WWW, TASCE, WAB.

ROBINSON, STILLMAN WILLIAMS; b. South Reading, VT, March 6, 1838; d. Oct. 31, 1910; f. Ebenezer, Jr.; m. Adeline Williams (Childs); w. Mary Elizabeth Holden; w. 2nd Mary Haines; c. Ekka M., Erdis G., Zella. Attended district school. Bound out to farmer until 1855. Apprentice in a machine shop, 1855-59. Entered the Univ. of Michigan in Jan. 1861, grad. with degree of civ. engr., 1863. Worked as an instrument maker during school. Invented machine for grading thermometers. Asst. engr. on US Lake survey, 1863-66. Instructor in civ. engr. at Univ. of Michigan, 1866-67. Asst. prof. of mining engr. and geodosy, 1867-70. Prof. of mech. engr. and physics at Univ. of Illinois, 1870-78. Prof. at Ohio State Univ., 1878-95, prof. emeritus in 1899. Inspector of RRs for Ohio, 1880-84. Inspector of RRs and bridges for RR commission of Ohio, 1880-84. Consulting engr. for Lick Telescope and mountings, 1887. Consulting engr. for Atchinson, Topeka and Santa Fe RR, 1887-90. Three awards granted on inventions at Centennial Exhibition of 1876 and one at Columbian Exhibition of 1893. Hon. Sc.D. from Ohio State Univ., 1896. Granted about 40 patents on several inventions, including boot and shoe nailing machines, 1882-89, improved Pitot tube instrument for measuring flow of gases and liquids in pipes and from gas wells and open streams, 1892, transmission-dynameter, 1898, Robinson Delmers Hypodermic syringe, 1898, Robinson Hitchcock automatic air brake mechanism, 1899, angle shaft coupling, 1903, and a lens grinding machine, 1911. Wrote three volumes in Van Nostrand's *Science* series and an enlargement and revision of another, 1876-84. Mem: ASCE, Am. Soc. of Mech. Engrs., Soc. of Naval Architects and Marine Mech., Am. Assn. for the Advancement of Science, and Soc. for the Promotion of Engr. Education (founder, 1890). Publ: *Treatise on the Teeth of Gear Wheels with the Theory and Use of the Robinson Templet Odontograph* (1876), *Railroad Economics* (1882), *Strength of Wrought Iron Bridge Members* (1882), *Compound Steam Pumping Engines* (1884), *Analytical and Graphical Treatment, Principles of Mechanism* (1896), co-authored "Red Rock Cantilever Bridge" (Thomas Fitch Rowland Prize 1892), and other articles on engr. and scientific subjects. Refs: BDNA, CAB, WWW, DAB, TASCE, Lamb, WAB.

RODD, THOMAS; b. London, England, June 13, 1849; d. Pittsburgh, PA, Aug. 3, 1929; f. Horatio; m. Anne (Theobald); w. Mary Watson Herron; c. William H., Thomas Jr., Mary Herron. Entered the US Navy in 1862, served until the end of the Civil War. Served as ship's writer, acting mate and capt.'s clerk. Last service on bd. the US steamer, *Galena*. Served under Capt. Francis Wells who recommended his appt. to the US Naval Academy. Cadet midshipman at Annapolis, 1865. Passed all examinations, resigned Feb. 1869 before graduation and the final exam. Rodman for Jesse Lightfoot, engineer and surveyor of Germantown, PA. Asst. under Strickland Kneass, chief engr. and surveyor with the city of Philadelphia 1869-72. Rodman for Pennsylvania RR Co., May 20, 1872-March 1, 1873, asst. engr. on const., March 1, 1873-July 1, 1877, principal asst. engr. July 1, 1877-July 1, 1889. Chief engr. of the Pennsylvania RR Co., July 1, 1889-Jan. 1, 1901. Chief engr. of the Pennsylvania lines west of Pittsburgh, PA, Jan. 1, 1901-July 1, 1918. Corporate engr. of the Pennsylvania lines west of Pittsburgh, July 1, 1918-June 30, 1919. Consulting engr. of the Pittsburgh, Ft. Wayne & Chicago RR Co., 1905-Jan. 1927. Chief engr. of the Chicago Union Station Co., Nov. 24 1913-May 1, 1919. Supervised erection of terminal at Chicago, Ill. Consulting engr. for Chicago Union Station Co., May 1, 1919-Dec. 31, 1925. Engr. and architect of const. of the Union Station for the Indianapolis Union RR Co. at Indianapolis, MN, 1887. Rebuilt the Pennsylvania RR after the flood of Johnstown, PA in 1889. Designed and const. structures including, the Westinghouse Electric and Manufacturing Co. Plant and Westinghouse Machine Co. works at East Pittsburgh, PA, Union Switch and Signal Co.'s Works at Swissville, PA and the plant of the British Westinghouse Electric and Manufacturing Co. at Trafford Park, Manchester, England. Investigated and made report on ore and coal handling facilities of Pennsylvania RR Co. at various points on Lake Erie, 1884-85. Mem: ASCE, Engrs.' Soc. of Western Pennsylvania (charter mem., dir., vice pres., 1881-2), Am. Railway Engr. Assn. (charter mem., dir., 1899-1903), Univ. Club of Chicago and a number of leisure clubs. Rfs: WWE, WAB, TASCE, WWW.

ROSEWATER, ANDREW; b. Bohemia, Oct. 31, 1848; d. 1909; w. Frances Meinrath. Educated in common and high schools of Cleveland, OH. Flagman of engr. corps, Union Pacific RR explorations and surveys, 1864. Occupied other positions in the same road. Asst. city engr. of Omaha, NB, 1868-70, city engr., 1870-75. Mgr. and interim editor of the Omaha *Bee*, 1876-77. Engr. in charge of const., Omaha & Nebraska RR, 1878-80. Resident engr. of the Omaha Water Works Co., 1880-81. City engr. of Omaha, 18881-87. Consulting and designing engr. of sewerage for 25 cities. Pres. of electric subway commission of Washington, DC, 1891-92. Consulting engr. for cities in CO and SD. City engr. of Omaha and pres. of the Bd. of Public Works, 1897-. Publ: "Report of the Electric Commission of the District of Columbia to the President" (1891). Refs: WWW.

ROTCH, WILLIAM; b. New Bedford MA, July 22, 1844; d. Aug. 14, 1925; f. William J; m. Emily (Morgan); w. Mary Rotch Eliot; c. Edith Eliot, Charles Morgan, Clara. Educated at the Friends Academy of New Bedford. grad. from Harvard Univ. in 1865, with degree of B.A. Entered the Ecole Centrale des Arts et Manufactures in Paris. Grad. in 1869 with degree in civ. engr. Asst. engr. of the Fall River, MA Water Works, 1871-74. Served as chief engr., 1874-80. Consulting engr. and purchasing agent of Mexico Central RR Co., Sonora RR Co., Atlantic and Pacific RR Co. and California Southern RR Co., 1880-82. Consulting engr. of the Cleveland, Canton and Southern RR Co., Lakeside and RR Co., New Bedford and Fall River RR Co., and other RR Companies in OH and MA, 1884-91. Engr. and mem. of commission which est. boundary line between RI and MA, 1881-83. Mem. of Re-organization Committee of the Atchinson, Topeka & Santa Fe RR, 1894-95. Dir. of Atchinson, Topeka and Santa Fe, 1895-1900. Dir. of Mexico Central RR Co., 1880-1900. In charge of 1st train which ran from Boston to Mexico City, April 1884. Pres. of Tremont Electric Lighting Co. and consolidated Electric Manufacturing Co. and a dir. of its successor, Walker Co. which installed the complete electric plant in the Waldorf-Astoria Hotel in NYC. Pres. of Federal Wharf and Storage Co., 1913-1922. Vice pres. of State Wharf and Storage Co., 1918-1922. Pres. of Fibre Manufacturing Co., 1918-22. pres. of RR Wharf and Storage Co., Denbigh Mining Corp., Terrible Dunderberg Mining and Powder Co., vice pres. of Bonanza Development Co., treas. of the Broadway Storage Co., trustee and treas. of Monquitt Real Estate Trust, managing dir. of Infants Hospital and the Adams Nervine Hospital. 1st pres. of Union for Good Works in New Bedford, 1870. Mem: ASCE, Alliance Francaise of Boston and Cambridge, MA (pres.), Massachusetts Soc. of Colonial Wars (governor), Boston Soc. of Architects (hon. mem.), Assn. des Anciens Eleves de l'Ecole Centrale of Paris, Bostonian Soc., Somerset and Harvard Clubs of Boston and Wamsutta Club of New Bedford. Refs: TASCE, WAB, WWW, WWE.

ST. JOHN, ISAAC MONROE; b. Augusta, GA, Nov. 19, 1827; d. White Sulfer Springs, WV, April 7, 1880; f. Isaac Richards; m. Abby Richardson (Munroe); w. Ella Carrington; c. six. Moved with family to NYC. Grad. from Yale with degree of A.B. in 1845, and A.M. in 1848. Studied law in NYC. Asst. editor of Baltimore *Patriot* at Baltimore MD, 1847-48. Civ. Engr. on Baltimore & Ohio RR until 1855. Supt. const. divisions of Blue Ridge RR in GA, 1855-61. Private in Ft. Hill Guards, SC state troops, Feb. 1861. Transferred to engr. duty, April 1861. Engr. in chief of confederate forces on the peninsula under Gen. John B. Magruder. Promoted maj., April 18, 1862. Chief of mining and nitre bureau corps in May 1862. Rose to rank of brig. gen., attaining the position of commissary gen. of Confederate States Army, Feb. 16, 1865. Chief engr. of Louisville, Cincinnati & Lexington RR, 1866-69. City engr. of Louisville KY, made 1st topographical map and est. sewer system, 1870-71. Consulting engr. of Chesapeake & Ohio RR, 1871. Chief engr. of Lexington & Big Sandy RR, 1871-80. Publ: "Resources of the Confederacy" (March 1877). Refs: Lamb, DAB, TASCE, BDNA, CAB, WWW.

SAYRE, ROBERT HEYSHAM; b. Columbia County, PA, Oct. 13, 1824; d. South Bethlehem, PA, Jan. 5, 1907; f. William H.; m. Eliza (Kent); w. Mary Evelyn Smith; w. 2nd Mary (Bradford) Brodhead; w. 3rd Helen Augusta (Packer) Rathburn; w. 4th Martha Finley Nevin; c. nine from 1st marriage, three from the fourth. Educated in common schools, then studied civ. engr. under James Nowlin. Employed on enlargement of Morris Canal and joined engr. corps under Andrew A. Douglas, 1840. Assigned to make surveys and const. RR between coal mines at Summit Hill and Mauch Chunk on Lehigh Canal, 1841-43. Built switchback RR and inclined planes into Panther Creek Valley Coal field. Placed in charge of all RRs and inclined planes of the co. and transportation of coal from the mines to Mauch Chunk. Directed development of the co. mines and erection and installation of equipment. Employed with Delaware, Lehigh, Schuylkill & Susquehannah RR and its successor, the Lehigh Valley System, May 1852-98. Exception of 1882-85 when employed as pres. and chief engr. of South Pennsylvania RR. Positions of chief engr. of Delaware, Lehigh, Schulykill & Susquehannah RR, May 1852-55, gen. supt. 1855-82, second vice pres., asst. to the pres., and vice pres. Const. rail outlet to tidewater. Built feeder lines from western coal fields, 1852-82, the Easton & Amboy RR, 1872-75, and extensions to Buffalo and Jersey city, 1885-98. Introduced iron bridges to replace wooden structures on line of the Lehigh Valley RR, 1857, use of steel rails, 1864, the fish bar truck joint and steel-tired driving wheels and steel fireboxes for locomotives. Promoter of Bethlehem Iron Co., dir., 1862, gen. mgr., 1886, vice pres., 1891. Promoter of Pioneer Mining and Manufacturing Co. of Alabama. Dir. of the Nescopec, the Upper Lehigh and the Wilmore Coal Co. in PA, South Bethlehem Gas and Water Co. and the E.P. Wilbur Trust Co. of South Bethlehem, Lehigh Salt Mining Co. of NY and other corporations. Treas.

of Bd. of Diocesan Missions of central PA, from organization of the diocese until 1890. Elected 3rd vice pres. when Philadelphia & Reading RR Co. leased Lehigh Valley RR in Feb., 1892-May 1893, retained 2nd vice pres. of Lehigh Valley RR. Chairmen of Bd. of Trustees and of Executive Committee of Lehigh Univ. Refs: DAB, WWW, WAB.

CHARLES CONRAD SCHNEIDER (1843-1916)
PRESIDENT OF ASCE, 1905

SCHNEIDER, CHARLES CONRAD; b. Apolda, Saxony, April 24, 1843; d. Philadelphia, PA, Jan. 8, 1916; f. Julius; m. Emilie (Bengel); w. Catherine Clyde Winters; c. one son, and Helen. Attended schools in Apolda. Apprentice in a machine shop, 1864-67. Attended Royal School of Technology in Chemnitz, Saxony, grad. 1864. Practiced as mech. engr., 1864-67. Came to the US, 1867. Draftsman with Rogers Locomotive Works at Paterson, NJ, 1867-70. Asst. engr. with Michigan Brodge and Const. Co. of Detroit, MI, 1871-73. In charge of engrs. office of Erie RR Co. in NYC under chief engr., Octave Chanute, 1873-75. Appt. by bd. of engrs. known as the Steinway Commission to prepare plans for a RR and highway bridge across East River to connect Long Island with NYC, 1876-77. Designer, Delaware Bridge Co. of NY, May 1877-July 1878. In charge of design and const. of bridges at Edge Moor, including Rockville Bridge over the Susquehanna River near Harrisburg, PA, on the Pennsylvania RR, Cohoes Bridge on the Delaware and Hudson RR, and several smaller ones. Opened office in NYC for design of bridge and structural work, Aug. 1, 1878-86. Consultant for Canadian Pacific RR, built Fraser River Bridge, 1882, and Stony

Creek Viaduct. Built Mareut Gulch Viaduct on Northern Pacific RR. Built Niagara Cantilever Bridge for Grand Trunk RR, 1883. Worked on inside finishing of the Statue of Liberty. Mem. of bd. of consulting engr. appt. by NY District RR Co., 1886. Chief engr. of bridge and const. dept. of Pencoyd Iron Works in Philadelphia, 1886-1900. Vice pres. in charge of engr., American Bridge Co., 1900-1903. Consulting engr. specializing in bridges and structural steel work, 1903. Consulting engr. of Baltimore & Ohio RR. Commissioned by the Imperial Government RRs of Japan, to prepare large set of standard plans for bridges for Japanese RRs. Reported on causes of collapse of the Quebec Bridge for the Canadian Government, 1907-1908. Mem. of bd. of engrs. for rebuilding the Quebec bridge, 1911-16. Awarded ASCE's Rowland Prize for paper on const. of the Niagara Bridge, 1886 and the Norman Medal for a paper on the structural design of buildings, 1905, and the Norman Medal for paper on movable bridges, 1886. 1st prize for design of Washington Bridge over the Harlem River, NYC. Mem: ASCE (dir., 1887, 1898-1900, vice pres., 1902-1903, pres., 1905), Am. Railway Engr. Assn., Am. Soc. for Testing Materials, Verein Deutscher Ingenieure in Germany, Canadian Soc. of NY, and Engrs. Club of NY. Publ: *General Specifications for Railroad Bridges* (1886), *General Specifications for Highway Bridges* (1901), *General Specifications for Structural Steel Work in Buildings* (1905). Refs: WWW, WAB, TASCE.

SCHUYLER, JAMES DIX; b. Ithaca, NY, May 11, 1848; d. Santa Monica, CA, Sept. 13, 1912; f. Philip Church; m. Lucy M. (Dix); w. Mary Ingalls. Educated at Friends College, 1863-68. Asst. on location and const. of the Kansas Pacific RR in western KS and CO, 1869-73. Resident engr. on the Denver & Rio Grande RR. Division engr. of North Pacific Coast RR in California, 1873-74. Chief engr. of Stockton & Ione RR, 1874. Wrote for Stockton Daily Independent. Chief asst. state engr. of California, in charge of irrigation investigations in Great Central Valley, 1877-82. Chief engr. and gen. supt. of Sinaloa & Durango RR in Mexico, 1882-84. Contract work in San Francisco, built sea wall, 1884-85. Designed Sweetwater Dam, 1886-88. Consulting engr. to Lake Hemet Water Co., designed and const. Hemet Dam, highest masonry dam in western US at time of const., 1891-95. Consulting included projects for Monterey Water Works and Sewer Co. of Monterey, Mexico; Mexican Light and Power Co. for 4 large dams in Necaxa Valley, Puebla, Mexico; Vancouver Power Co. Ltd., Vancouver, British Columbia; Kobe Syndicate of Japan ofr Nuuanu Dam in Hawaii. Built water supply systems for Denver, Odgen, UT, and Los Angeles. Appt. one of three consulting engrs. to report on plans for Los Angeles aqueduct. On commission of consulting engrs. appt. to judge feasibility of the Gatun Dam and other structures for the Panama Canal. Mem: ASCE (vice pres. and dir.), Soc. of Civ. Engrs. of London, Am. Geographical Soc., Franklin Inst. of Philadelphia PA, and Technical Soc. of the Pacific Coast. Publ: "Recent Practice in Hydraulic-Fill Dam Construction" (1907), "The Construction of the Sweetwater Dam" (1888), *Reservoirs for Irrigation, Water Power, and Domestic Water Supply* (1901 and 1908), and other contributions to technical journals. Refs: WAB, DAB, WWW, CCI, TASCE, Schnitter.

SEARLES, WILLIAM HENRY; b. Cincinnati, OH, June 4, 1837; d. April 25, 1921; f. Asbury M.; m. Rachel (Mitchell); w. Mary Doolittle. Educated at private schools and Weslyan Univ., 1856-57. Entered RPI in Feb. 1857, grad. 1860. Asst. engr. of Marietta & Cincinnati RR, 1860-61. Asst. engr. of military defenses, dept. of the Ohio, 1861-62. Prof. of geodosy and road engr., RPI, 1862-64. Chief engr. of Middle Range RR, Michigan, 1864. Asst. engr. of Pittsburgh, Ft. Wayne & Chicago RR, 1865-66. Principal asst. engr. of Allegheny Valley RR, 1866-67. Chief engr. of Indiana North and South RR, 1870-71. Chief Engr. Corps No. 7, NY, West Shore and Chicago RR, 1872-73. Consulting engr. in NY, 1874-75. Division engr., NY state canals, 1876-78. Consulting engr. American Pier and Column Co., NY, 1879-80. Division engr. of NY West Shore and Buffalo RR, 1881-82. Chief engr. of the Williamsport and Clearfield RR, PA, 1883. Asst. chief engr. of B. C. Clearfield and S. W. RR, 1884. Civ. and consulting engr. in Cleveland, OH. Built Indiana North & South RR, West Point Tunnel at West Point, NY, RR lift bridge for B. C. Clearfield and S. W. RR over west branch of the canal, Jersey Shore, PA. Mem: ASCE, Cleveland Engr. Soc. (hon. mem., mgr., 1888-90, pres., 1890-91, secy., 1897). Publ: *Field Engineering* (1880), *The Railroad Spiral* (1882), and several papers in scientific journals. Refs: TASCE, RPI Bio, WWW.

SERRELL, EDWARD WELLMAN; b. London, England, Nov. 5, 1826; d. NYC, April 25, 1906; f. William; m. Anna (Boorn); w. Jane Pound; w. 2nd Marion Seaton Roorbach. Came to the US when he was 4 years old with his parents, who were US citizens. Educated at home and academy of the Gen. Soc. of Mechanics and Tradesmen. Grad. from Leggett and Guillaurdeau's Collegiate School, 1841. Worked as civ. engr. with his father and older brother. Engaged in RR and bridge design and const., 1845-61. Asst. engr. to commissioners of Erie RR, 1846-47. Asst. to Chief of Topographical Engrs., US Army. Asst. engr. of Panama Survey, 1848. With expedition that located the route between Aspinwall and Panama. In charge of surveys for Northern RR of New Hampshire. Engr. of Central RR of NJ. Planned and supervised the const. of suspension bridge over Niagara River at Lewiston, NY, 1850. Supt. const. of bridge at St. John, New Brunswick. Planned bridge over the St. Lawrence at Quebec. Involved with const. of Hoosac Tunnel, 1854 and Briston Bridge over the Avon River in England, 1855. Chief engr. of Union Pacific RR. Projected fortifications for protection of Washington, DC, 1861. Recruited and commanded 1st NY Engrs. with rank of lt. col., then col. Accompanied Sherman's expedition to Ft. Royal. Built works at Wales Cut, in front of Ft. Pulaski, on Morris Island and Folly Island. Asst. engr. of Dept. of the South. Chief engr. of the X Corps, US Army and of the Dept. of the South. Chief engr. and chief of staff of Army of the James. Devised and supervised const. of a battery in a swamp, "Swamp Angel". Chief of staff and of engrs., detailed on special scientific duty. Served in 126 actions at the end of his army career, brevetted brig. gen. on March 13, 1865. Mustered out on Feb. 13, 1865. Invented improvements in armor plate, impromptu gun carriages, and development of electric coast defenses while in the army. Reported for RRs and canals. Served as consulting engr. for corporations, including American Isthmus Ship Canal Co. Reported on location of suspension bridge to cross the Hudson River, 1868. Promoted const. of bridge at this location in connection with RR from the PA coalfields to New England, never carried to completion. Projected an Isthmian canal from the Atlantic to the Pacific at sea level and without locks, nearer to NY than the Panama or Nicaragua, 1902-1903. Inventions included sand boxes for locomotives, long wire for telegraphs and bridges, and test boxes for underground telegraph wires. Refs: DAB, Lamb, BDNA, CAB.

SHANKLAND, EDWARD CLAPP; b. Pittsburgh, PA, Aug. 2, 1854; d. June 3, 1924; f. Edward Russell; m. Elizabeth F. (Clapp); w. Harriet S. Graham. Educated in public schools. Attended Iowa State Agricultural College at Ames, IA. Went to Cornell College at Mt. Vernon, IA. Entered RPI, Sept. 1875, grad. with degree of civ. engr., 1878. Entered US Army. Engaged on improvements of the Missouri and Mississippi Rivers at Nebraska City, NE, on the Missouri River Survey at Kansas City, MO and at Carrolton, IA, Aug. 1878-April 1883. In charge of improvements at Lexington, MO for two years. Principal asst. engr. on Leavenworth Division. Resigned from govt. employment, 1883. Asst. engr. with Wrought Iron Bridge Co. of Canton, OH. Entered office of Burnham & Root, architects in Chicago, IL, 1888. Opened office under his own name. Engr. of const. of World's Columbian Exposition. Designed roofs of the machinery hall and the building for manufactures and liberal arts. Mem. of Harbor Survey Commission, Chicago, 1911-16. Hon. degree of M.A. from Cornell College, IA, 1904. Mem: ASCE, Inst. of Civ. Engrs. of Great Britain, Rensselaer Soc. of Engrs. and Western Soc. of Engrs. Refs: WAB, RPI Bio, WWW.

SHUNK, WILLIAM FINDLAY; b. Harrisburg, PA, Sept. 6, 1830; d. 1907; f. Francis R.; m. Jane Findlay; w. Gertrude Wyeth. Educated in public schools and at Harrisburg Academy. Attended Dickinson College. Midshipman in US Navy, 1846-50. Worked with the Pennsylvania RR, 1851-56. In the US Coast Survey, 1856. Asst. engr. of the Lewisburg & Spruce Creek RR, 1856-57. Clerk in the State Dept. of Washington, 1861-65. Rock Island US Govt. Surveys, 1865-66. Asst. engr. of the Dutchess & Columbia RR, 1867. Chief engr. of Connecticut Western RR, 1868-74. Chief engr. of the Metropolitan Elevated RR, NY during const., and of the Manhattan Elevated RR after const. of the roads, 1876-82. Chief engr. of South Pennsylvania RR, 1882-85. Chief engr. Kings Co. Elevated RR, Brooklyn, 1887-89. Engr. in charge of intercontinental RR surveys, 1890-92. Chief engr., Guayaquil & Quito RR, Equador, 1898-1902. Publ: *Shunk on Railway Curves* (1854), The Field Engineer (1879-1903). Refs: WWW.

SIBERT, WILLIAM LUTHER; b. Gadsden, AL, Oct. 12, 1860; d. Bowling Green, KY, Oct. 16, 1935; f. William J.; m. Marietta (Ward); w. Mary Margaret Cummings; w. 2nd Juliette Roberts; w. 3rd Evelyn Clyne Bairnsfather; c. William, Franklin, Harold, Edwin, Martin, and Mary. Educated in rural schools until he was 14. Attended Univ. of Alabama, 1878-80. Appt. to US Military Academy, grad. seventh in class of 1884. Appt. 2nd lt. of engrs., June 15, 1884. Grad. from Engr. School of Application, 1887. In charge of engr., River and Harbor Districts, headquarters in Louisville, KY, July 1887-April 1888. 1st lt., April 7, 1888. In charge of local improvement works on Green and Barren Rivers near Bowling Green, KY, April 7, 1888-Aug. 1892. In const. of Sault Ste. Marie Canal connecting Great Lakes, 1892-94. At waterways in Great Lakes Region. Command of River and Harbor District at Little Rock, AL, Aug. 1894-Sept. 1896. Capt., March 31, 1896. Returned to engr. school of application as instructor in civ. engr., Sept. 1898. Commanded co. of engr. troops in Manila,

GATUN LOCKS UNDER CONSTRUCTION
BUILT BY MAJ. WILLIAM L. SIBERT BEGINNING IN 1908

July 5, 1899. Appt. chief engr. of VIII Army Corps and mem. of staff of Gen. Elwell S. Otis. Chief engr. of Dept. of the Pacific as well as commanding the engr. batallion, Sept. 2-Nov. 25, 1899. With Gen. Schwan's Brigade against insurgents in Northern Luzon, Oct. 7-14, 1899. With campaign through provinces of Cavite, Laguna, Batangas, and Tabayas in Southern Luzon, Jan. 2-Feb. 5, 1900. Returned to US on the Meade, May 5, 1900. District engr. at Louisville, KY, July, 1900. Transferred to District engr. of Pittsburgh, Dec. 1901-March 1907. Maj., April 23, 1904. Mem. of Isthmian Canal Commission, March 1907-April 1, 1914. Lt col., Sept. 21, 1909. Built Gatun Locks and Dam, Panama Canal, west breakwater, Colon Harbor, excavated channel from Gatun to the Atlantic Ocean beginning June 30, 1908. Served as chair of bd. of engrs. on flood prevention problem under the Am. National Red Cross and Chinese govt., Huai River Valley, China, June 11-Oct. 15, 1914. Division engr., Central River and Harbor Division, and district engr., 2nd Cincinnati District. Mem. of bd. on const. or modification of locks and dams on Ohio River. Mem. of bd. on flood conditions in valley of Ohio River and in drainage area of Lake Erie. Brig. gen., US Army, received thanks of Congress by act approved March 4, 1915. In command of Pacific Coast Artillery District, March 4, 1915-May 15, 1917. Assigned as cmdr., 1st Division, Am. troops in France, under Maj. Gen. Pershing, June 8, 1917. Maj. gen., June 28, 1917. In action with Division in the Luneville Sector, Oct. 21-Nov. 20, 1917. Cmdr. of Southeastern Dept. at

Charleston, SC, Dec. 4, 1917-May 16, 1918. Organized and directed Chemical Warfare Service, US Army, May 1918-March 1, 1920. In command of 5th Division and Camp Gordon, GA. Retired, April 4, 1920. Chair and chief engr. of Alabama State Docks Commission, Nov. 26, 1923. Chair of Boulder Dam Commission, 1928. Retired from engr. to his farm. Cmdr., French Legion of Honor. Awarded Distinguished Service Medal for his organization and administration of Chemical Warfare Service. LL.D. from Univ. of Nebraska, 1919. D.Engr. from the Univ. of Nebraska, 1919. Mem: Am. Assn. of Port Authorities (pres., 1929-30). Refs: WWW, DAB, TASCE, WAB, WWE.

SICKELS, FREDERICK ELLSWORTH; b. near Camden, NJ, Sept. 20, 1819; d. Kansas City, MO, March 8, 1895; f. John; m. Hester (Ellsworth); w. Ranane Shreeves; c. five. Rodman on Harlem RR. Apprentice under James P. Allaire of NY. Patented 1st successful drop cut-off for steam engines in US, 1841, patented, 1842. Built it into engine of steamboat on Long Island Sound, the *Champion*. Applied to many steamers on that sound and the Hudson River, the last was the *North America*. Adopted by US Navy, employed on the *Waterwitch* in 1847, on the *Powhatan* in 1848 and others. Patent pended for steam steering apparatus, 1849-60. Completed const. of full size steam steering unit, 1854. Installed equipment on steamer Augusta, 1858. Patented July 17, 1860. No purchaser for invention, made trip to Europe, found no buyer, 1860-67. Const. Omaha Bridge over Missouri River. Consulting engr., National Water Works Co. of NY, 1890. Chief engr. of operations at Kansas City, MO, 1891. Mem: ASCE. Refs: WWW, DAB, TASCE.

SIDELL, WILLIAM HENRY; b. NYC, Aug. 21, 1810; d. NYC, July 1, 1873; f. John Sidell. Appt. by NY state to US Military Academy on July 1, 1829, grad. 6th in class of 1833. Brevetted 2nd lt. of 1st Artillery, US Army, 1833. Resigned, Oct. 1, 1833. City surveyor in NYC, asst. engr. on Croton Aqueduct, division engr. of Long Island RR, asst. engr. on dry docks on NY Harbor. Engr. on US Hydrographic Survey of delta of the Mississippi, 1837-39. Engr. with various RRs in NY and MA, 1839-46. Captain during war with Mexico, regiment never mustered in. With Isthmus of Panama RR, 1846-49, last two years as chief engr. In US service, exploring for a RR route from Mississippi to the Pacific. Surveyed RR route across isthmus of Tehuantepec in Mexico for James Eads, 1851-52. Chief engr. of RRs in IL and MO until 1858. Chief engr. of Louisiana Tehuantepec Co. to survey transisthmian RR route. Commissioned maj. of 15th Infantry in regular army and assigned to recruiting duty in regular army in KY and TN, May 14, 1861. Appt. acting asst. adjutant gen. of Dept. of the Cumberland, 1862. Acting asst. provost marshal gen. for KY and chief mustering and disbursing officer, Louisville, KY, 1863-65. Commissioned lt. col., 10th Infantry, May 6, 1864. Brevetted col. and brig. gen., May 13, 1865. On frontier duty in ND, SD and KS. Retired, Dec. 15, 1870. Refs: WWW, DAB, CAB.

SIMPSON, JAMES HERVEY; b. New Brunswick, NJ, March 9, 1813; d. St. Paul, MN, March 2, 1883; f. John Neely; m. Mary (Brunson); w. Jane Champlin; w. 2nd Elizabeth Sophia (Borup) Champlin; c. four daughters. Educated in common schools. Entered US Military Academy, grad. 1832. Commissioned in artillery. Aide to Gen. Henry Lawrence Eustis during Seminole War, 1837-38. In action at Locha Hatchee. Transferred to

Topographical Engrs., 1838. Engaged in engr. projects in the east and south, 1838-60. In charge of exploration of route from Ft. Smith, AR to Santa Fe, NM. Chief topographical engr., Dept. of New Mexico. Reconnnoitered route from Santa Fe to the Navajo Indian country. Capt., March 3, 1853. Five years on road const. in MN and two years on coast survey duty. Accompanied Utah Expedition and submitted report on new route from Salt Lake City to Pacific Coast, 1858. Rank of Maj. Chief topographical engr. for a few months, Dept. of the Sheneandoah. Col., 4th NJ Volunteers. Engaged in peninsular campaign. Engaged at Westpoint, VA and at Gaine's Mill where taken prisoner. Later exchanged. Chief topographical engr. and chief engr., Dept. of the Ohio, 1862-63. In gen. charge of fortifications and engr. projects in KY, 1863-65. Brevetted col. and brig. gen., March 13, 1865. Chief engr., Dept. of the Interior, 1865-67. Gen. direction and inspection of Union Pacific RR and all govt. wagon roads. Col. of engrs., March 7, 1867. Engaged in road const., river and harbor improvements and lighthouse supervision in the south and middle west. Retired at own request, March 31, 1880. Publ: *Report from the Secretary of War Communicating the Report and Map of the Route from Ft. Smith, Arkansas to Santa Fe, New Mexico* (1850), *Journal of a Military Reconnaissance from Santa Fe, New Mexico to the Navajo Country* (1852), *The Shortest Route to California* (1869), *Coronado's March in Search of the Seven Cities of Cibola* (1871), *Exploration Across the Great Basin of the Territory of Utah* (1876), and other reports in explorations of the west. Refs: DAB, WWW, CAB, DR.

SLIFER, HIRAM JOSEPH; b. Colmar, PA, Oct. 12, 1857; d. Feb. 3, 1919; f. John; m. Lydia (Huttel); w. Mary A. Beatty. Grad. from course in mining and civ. engr. at Polytechnic College of Pennsylvania, 1876. Engaged as rodman, leveler and transitman for the Mexican National Const. Co., 1879-82. Asst. engr. for the Philadelphia Division of the Pennsylvania RR, 1882-91. Principal asst. engr. of Milwaukee, West Shore and Western RR, 1891-93. Division engr. of the Ashland Division C. & N. W. RR, at Kaukauna, WI, 1893-97. Division engr. of second track in IA and supt. of the Iowa Division of the same road, 1897-1902. Gen. supt., Eastern District of C. R.I. & P. RR, 1902-1903. Central District, 1903-1905. Steam RR expert and business mgr., const. dept. J.G. White & Co., NY, 1905-1907. Gen. mgr., Panama RR & SS lines at Colon, Panama, 1907-1909. Gen. mgr., C. G. W. RR, 1909-12. Commissioned lt. col. 21st Engrs. and with A. E. F. in France. Mem: Mason. Refs: WWW.

SLOAN, ROBERT IMLAY; b. Feb. 23, 1837; d. March 3, 1901; f. William Henry; m. Caroline (Imlay); w. Miss Rupert; c. Robert I. Jr. and one other child. Rodman on Flemington Branch of the Belvidere and Delaware RR, also leveler, transitman and asst. engr., 1852-55. Entered RPI, fall 1855, grad. in 1859. Asst. engr. on St. Joseph & Topeka RR and on the St. Joseph & Marysville RR in KS, 1866. Quartermaster's Dept. of Army of the Potomac in VA, 1861-64. Surveyor of oil tracks in Venango County, PA, 1865. Asst. engr. on Atlantic & Great Western RR and on surveys for improvement of Rock River and the Mississippi River under US govt., 1865. City engr. of Trenton, NJ for four years. Principal asst. engr. on surveys of Metropolitan Elevated RR of NYC, Feb. 1, 1876. Designed foundations and track system of Metropolitan Elevated RR on West Side, NYC. Appt. chief engr. on Aug. 1, 1881-March 1, 1890. Chief engr. of Chicago South Side Elevated RR, March 1, 1890. Engr. for J. B. and J. M.

Cormell, in charge of erection of Waldorf-Astoria Hotel in NYC. Mem. of firm, Hall & Co. of NY. Mem: ASCE. Refs: TASCE, RPI Bio.

SMITH, CHARLES SHALER; b. Pittsburgh, PA, Jan. 16, 1836; d. St. Louis, MO, Dec. 19, 1886; f. Frederick Rose; m. Mary Anne (Shaler); w. Mary Gordon Gairdner; c. several. Attended a private school in Pittsburgh. Rodman on Mine Hill & Schuylkill Haven RR. Asst. engr. of Louisville & Nashville RR, 1855. Resident engr. on Memphis Bridge, 1856. Supt. of track and bridge const. of Memphis Division, 1859. Chief engr. of buildings and bridges, Wilmington, Charlotte & Rutherford RR, 1860-61. Served as capt. of engrs., Confederate Army in the Civil War. Built powder mill in Augusta District, 1865. Const. Catawba and Conagree Bridges on Charlotte & South Carolina RR. Partner with Benjamin and Charles Latrobe in firm of Smith, Latrobe and Co., 1866, became Baltimore Bridge Co., 1869. Pres. and chief engr., lived in St. Charles, MO, 1868-71. Built a series of iron trestles on Louisville, Cincinnati & Lexington RR, and Elizabethton & Paducah RR, 1868-69. Built KY River bridge for Cincinnati Southern RR, 1876-77. Cantilever became dominant type for long span const. as a result of work including Mississippi River bridge near St. Paul, and Canadian Pacific RR Bridge across St. Lawrence River. Mem: ASCE (dir., 1873). Publ: *Comparative Analysis of the Fink, Murphy, Bollman and Triangular Trusses* (1865), "Draw Spans and Their Turntables" (1874), "Proportions of Eye Bar Heads and Pins as Determined by Experiment" (1877), "Wind Pressure Upon Bridges" (1881). Refs: WWW, WAB, DAB, CAB.

HIGH BRIDGE, WILMORE, KY
BUILT BETWEEN 1876 AND 1877 BY CHARLES SHALER SMITH
TALLEST (255 FT.) BRIDGE IN THE WORLD AT THE TIME

SMITH, GUSTAVUS WOODSON; b. Scott County, KY, Jan. 1, 1822; d. NYC, June 25, 1896; f. Byrd; m. Sarah Hatcher (Woodson); w. Lucretia Bassett; c, none. Attended county schools. Grad. from US Military Academy in 1842. Asst. engr. in const. of Ft. Trumbell and Battery Griswold in New London Harbor, CT, 1842-44. Asst. prof. of engr. at US Military Academy, 1844-1846. Promoted 2nd lt., 1845. Commanded co. of sappers, miners and pontoniers during war with Mexico. Brevetted 1st lt., April 18, 1847, for Cerro Gordo. Capt., Aug. 20, 1847 for Contreras, Mexico. In the battle of Churubusco, Aug. 20, 1847. Const. battery, Chapultepec and in assault and capture of Mexico City, Sept. 14, 1847. Principal asst. prof. of engr. and art of war at the US Military Academy, 1849-54. Promoted 1st lt. 1853. Resigned commission 1854. Supt. extension of US Treasury Building at Washington, DC, 1855. Supt. repairs of branch mint and const. of Marine Hospital, New Orleans, 1855-56. Chief engr. of Trenton Iron Works, NJ, 1856-57. Deputy st. commissioner of NYC, April-Nov. 1858. Chief of St. Dept., 1858-Sept. 1861. Mem. of bd. to revise the program of instruction at US Military Academy, 1860. Joined Confederate Army at start of the Civil War. Appt. brig. gen. from KY. Commanded 4th Brigade, Army of the Shenandoah at 1st battle of Bull Run, July 21, 1861. Promoted maj. gen. in 1861. Assigned to command 2nd Division, Army of the Potomac, Oct. 22, 1861, of 2nd Corps, Army of Northern Virginia, March 14, 1862, of reserve, Army of Northern Virginia, April 18, 1862. Commanded the left wing of Confederate Army at battle of Seven Pines 1862. In command of Army of Northern Virginia when Gen. J. E. Johnston was wounded. Assigned to command of division in Army of Northern Virginia, Aug, 1862. In command of Dept. of North Carolina and Southern Virginia, Sept. 1862. Acting Secy. of War of Confederate States govt., Nov. 1862. Placed in command at Goldsboro, NC, Dec. 1862. Posted on Gen. Cheatham's right in battle of Atlanta. Stationed at Macon, GA where he resisted Sherman's advance, fighting battle of Griswoldsville and badly defeated by Charles R. Woods' division. Surrendered at Macon, GA, April 20, 1865 and held as prisoner of war. Moved to Chatanooga, TN, 1866. Supt. of Southwest Iron works, 1866-69. Insurance commissioner for KY, 1870-76. Moved to NYC, 1876. Publ: *Notes on Life Insurance* (1870), *Confederate War Papers* (1884), *Battle of Seven Pines* (1891), *Gen. J. E. Johnston and Gen. G. T. Beauregard at Manassas* (1892), *Company "A" Corps of Engrs., U.S.A. in the Mexican War* (1896). Refs: Lamb, DR, BDNA, WAB, CAB, DAB, WWW.

SMITH, HAMILTON; b. near Louisville, KY, July 5, 1840; d. Durham, NH, July 4, 1900; f. Hamilton Sr.; m. Martha (Hall); w. Mrs. Charles (Jennings) Congreve. Lived with grandfather while attending schools of Durham, NH. Studied engr. in father's coal mines at Cannelton, IN, 1858. Became chief of engr. and accountant's depts. Engaged in developing collieries in Kentucky and Indiana and superintending the RRs and machinery which were connected. Left for CA, 1869. Consulting engr. of Triunfo Mine. Engr. and mgr. of North Bloomfield and Milton Gold Mines in Nevada County. Designed and const. large dams, pipe lines and other works, authority on hydraulic mining in CA. Connected with other mines on the Pacific Coast and in Mexico. Active in efforts to cheapen manufacture of high explosives and in est. Vulcan Powder Works. Became consulting mining engr. for Rothschild Interests. Consulting engr. for El Callo Mine in Venezuela, 1881-85. Designed and supt. const. of machinery and appliances. Founded Exploration Company, Ltd. with Edmund

De Crano, 1886. Visited South Africa in 1892 and 1895. Took part in formation of Consolidated Deep Levels, Transvaal and General Assn., and other enterprises. Also connected with Alaska Treadwell, Alaska United, Alaska Mexican Gold Mining, Anaconda Copper Mining and other co. Partner died, 1895. New firm with H. C. Perkins, headquarters in NYC, active in promotion of Central London RR, 1890. Mem: ASCE, Am. Inst. of Mining Engrs. and Inst. of Civ. Engrs. of Great Britain. Publ: *Hydraulics: The Flow of Water through Orifices, over Weirs, and through Open Conduits and Pipes* (1886), "Costs of Mining and Milling Free Gold Ores," "The Flow of Water Through Pipes," "Water Power with High Pressures and Wrought Iron Water Pipe," and "Temperature of Water at Various Depths in Lakes and Oceans". Refs: TASCE, WWW, DAB.

SWAIN, GEORGE FILLMORE; b. San Francisco, CA, March 2, 1857; d. Holderness, NH, July 1, 1931; f. Robert Bunker; m. Clara Fillmore; w. Katherine Kendrick; w. 2nd Mary Hayden Lord; w. 3rd Mary Augusta Rand; c. Barbara, Clara. Attended military school, valedictorian of his class. Lived with relatives in Providence, RI after his father died in 1872. Studied with private tutor. Entered MIT, grad. in class of 1877 with degree of B.S. in civ. and topographical engr. Three years in Europe. Studied at Royal Polytechnicum in Berlin, 1877-80. Expert and special agent of the 10th US Census to report on water power in connection with manufacturing along Atlantic seaboard, 1880-84. Instructor in civ. engr. at MIT, 1881, asst. prof. 1883-86, assoc. prof. 1886-87, prof. 1887. Expert witness in failure of the Bussey Bridge in 1887. Expert engr. of RR Commission of MA, 1887-1914. Connected with NY, New Haven & Hartford RR, NY Central and Royal Commission on RRs and Transportation of Canada. Mem. of Boston Transit Commission, 1894-1918, chair for five years. Mem. of Inland Waterways Commission and National Conservation Commission. Appt. Gordon McKay Prof. of civ. engr. at Harvard Univ., 1909, emeritus, 1929. LL.D. from New York Univ., 1907, Univ. of California, 1918. Mem. of delegation of Am. engrs. to France, 1918. Mem. of Franco-American Engineering Commission, 1919. Mem. of bd. of judges to select names for Hall of Fame at New York Univ. Benjamin Garver Lamme Gold Medal from Soc. for Promotion of Engr. Education, 1st recipient, 1928. Mem: ASCE (hon. mem., dir., 1901-1903, vice pres., 1908-1909 and pres., 1913), Soc. of Arts at MIT (secy.), Technology Alumni Assn., MIT (secy.), Soc. for the Promotion of Engr. Education (pres.), Boston Soc. of Civ. Engrs. (pres. and hon. mem.), New England Water Works Assn., Am. Soc. of Mech. Engrs., Am. Inst. of Consulting Engrs., Am. Soc. for Testing Materials, Canadian Inst. of Mech., Inst. of Civ. Engrs. of Great Britain, Soc. of Engrs. of Hanover, Germany, Am. Railway Engr. Assn., New England RR Club, National Academy of Arts and Sciences, Am. Forestry Assn., St. Botolph, Technology and Commercial Clubs of Boston. Publ: *Water Power of the Southern Atlantic Water Shed* (1881), *Structural Engineering: Strength of Materials* (1924), *Structural Engineering: Fundamental Properties of Materials* (1924), *Structural Engineering: Stresses, Graphical Statics and Masonry* (1927) *Conservation of Water by Storage* (1915), *How to Study* (1917), *The Young Man and Civil Engineering* (1922). Refs: WWW, WWE, DAB, WAB, TASCE.

TALBOT, ARTHUR NEWALL; b. Cortland, IL, Oct. 21, 1857; d. Chicago, IL, April 3, 1942; f. Charles A.; m. Harriet (Newall); w. Virginia Mann Hammet; c. Kenneth, Mildred Virginia, Rachel Harriet, Dorothy Newell. Early education in Cortland and high school in Sycamore. Taught country school for two years. Entered Illinois Industrial Univ. at Urbana, grad. B.S. 1881, civ. engr., 1885. Asst. engr. on const. of Santa Fe RR and Denver & Rio Grande RR. Worked on location, const. and maintenance in CO, NM, KS and ID. Asst. prof. in engr. and mathematics at Univ. of Illinois. Prof. of municipal and sanitary engr., 1890-1926. Retired in 1926 and became prof. emeritus. Developed formulas for computing rates of maximum rainfall and size of culverts. Est. center for laboratory research on practical municipal problems, and est. Engr. Experiment Station at Univ. of Illinois, 1903. Mem. of bd. which determined the Galveston Texas Causeway. Served on bd. which gave preliminary report on location of San Francisco-Oakland Bay Bridge, 1927. Chair of special committee of the Am. Railway Engr. Assn. to determine stresses on RR track, 1914-41. Mem. of Bd. of Visitors to US Naval Academy, 1918-21. Mem. of Division of Engr. of National Research Council, 1919-20. Representative of ASCE on the Engr. Council. Mem. of Joint Committee on Concrete and Reinforced Concrete of ASCE and the Am. Soc. for Testing Materials, 1904. Reported on studies of const., mode of action and resistances of RR ties, ballast and roadbed under different loads at varying speeds. Received Chanute Medal in mech. engr. of Western Soc. of Engrs., 1908; Washington Award of the Western Soc. of Engrs., 1924; George Henderson Medal of Franklin Inst. for innovations in RR engr., 1924; Henry C. Turner Medal of Am. Concrete Inst. for work on reinforced concrete, 1928; Benjamin Garver Lamme Medal of Soc. for Promotion of engr. Education for achievements in engr. education, 1932; John Fritz Medal of the United Engr. Societies, 1937. Hon. doctorates from Univ. of Pennsylvania, 1915, Univ. of Michigan, 1916 and the Univ. of Illinois, 1931. Laboratory building at Univ. of Illinois named for him, 1938. Mem: Illinois Soc. of Engrs. (founder, pres., 1890-91), Soc. for Promotion of Engr. Education (pres., 1910-11), Am. Soc. for Testing Materials (hon. mem., pres., 1913-14), ASCE (hon. mem., pres., 1918), Am. Assn. for the Advancement of Science (vice pres., 1928), Am. Railway Engr. Assn., Am. Soc. of Mech. Engrs., Am. Public Works Assn., Am. Water Works Assn. (hon. mem.), and New England Water Works Assn., Am. Public Health Assn., Intl. Assn. for Testing Materials, Western Soc. of Engrs. (hon. mem.), Am. Concrete Inst. (hon. mem.), Inst. of Structural Engrs. (hon. mem.; Great Britain) and Soc. of Civ. Engrs. of London, Univ. and Engrs. Clubs of Chicago, Univ. Club of Urbana. Publ: *The Railway Transition Spiral* (1899), and about 400 other publications. Refs: DAB, WWW, WAB, WWE.

TATHAM, WILLIAM; b. Hutton-in-the-Forest, Cumberland, England, April 13, 1752; d. Feb. 22, 1819; f. Sanford; unmarried. Came to VA, 1769. Clerk in trading house of Carter & Trent on James River. Moved to Watauga settlement in TN country, employed in mercantile est. Clerk of Watauga Assoc. Drafted petition of inhabitants on the western waters praying for incorporation into govt. of NC, July 5, 1776. Fought in the Revolution, in the defense of Ft. Caswell on Watauga, July 1776, and at Yorktown, Oct. 1781. Wrote "History of the Western Country" in collaboration with Col. John Todd, 1780, but never printed. Mercantile venture in Philadelphia. Visited Havana, 1783. Moved to VA, clerk of council of state. Studied law, 1st under Samuel Hardy then under William R. Davie. Admitted to the bar, March 24, 1784. Delegate from

Robeson County in gen. assembly of NC in 1787. Elected lt. col. of militia. Organized a geographical dept. for VA. Returned to Tennessee country, practiced law, mapped the region and gathered material for its history, 1792. Visited Spain, 1796 in connection with affairs in the west, ordered to leave the country. Went to London and wrote articles for magazines and other publications on engr. and agricultural subjects. Supt. of const. of Wapping Docks in the Thames at London, 1801. Returned to US in 1805. Engaged in survey of coast from Cape Fear to Cape Hatteras. Offered collection of manuscript maps and historical data to Congress, 1806 and 1817, unsuccessful in sale. Proposal of 1806, was 1st to define functions of a national library for the US. 5 years as draftsman and geographer in Dept. of State at Washington, DC. Position in govt. arsenal on the James River, 1817. Died by deliberately stepping in front of a gun fired in salute. Publ: *A Topographical Analysis of the Commonwealth of Virginia from 1790-91, A Plan for Insulating the Metropolis by a Canal* (1797), *Remarks on Inland Canals* (1798), *The Political Economy of the Inland Navigation* (1799,) *An Historical and Practical Essay on the Culture and Commerce of Tobacco* (1800), *Auxiliary Remarks on an Essay on the Comparative Advantages of Oxen in Tillage* (1801), *National Irrigation* (1801), *Report on a View of Certain Impediments and Obstructions, in the Navigation and Conservancy of the River Thames* (1803). Refs: WWW, DR, DAB.

TERZAGHI, KARL; b. Prague, Bohemia, Austro-Hungary, Oct. 2, 1888; d. Winchester, MA, Oct. 25, 1963; f. Anton Terzaghi von Pontenuovo, military officer; m. ____ ; w. Olga Byloff; married 1916, divorced 1926; w. Ruth Dogget, married 1928; c. 1 son, 1 daughter. Graduated from Realschule, Prague, 1900; graduated from Technische Hochschule, Graz, Austria, 1904 with degreee in mech. engr. Doctor of Technical Sciences, Techniche Hochschule, Graz, Austria in reinforced concrete construction, 1912. Army service, 1905. Injured mountain climbing, 1906. Engr. for Adolph Baron Pittel, hydroelectric projects until 1908. Engineering surveys in Croatia until 1910. Dam studies in US under F. H. Newell until 1913. Served in Austro-Hungarian Army and Air Service, 1914-1916. Worked with von Karman and von Mises. Posted to Imperial School of Mech., Istanbul, Turkey and taught until 1918 under Prof. Philip Forchheimer. Accepted lectureship at Roberts College, Istanbul, and set up soil testing laboratory for earth pressures. Guest lecturer at MIT, 1925-1929. Prof. at Technical Univ. of Vienna, Austria, 1929-1938. 8 month lecture tour in US, 1936 President of the 1st Intl. Conference on Soil Mechanics and Foundation Engineering, 1936. Consultant to projects in Soviet Union, North Africa and Central Asia. Est. systematic settlement observations of buildings in Vienna. Left Austria when Nazi Germany took over and went to US via France and Great Britain. Accepted visiting professorship at Harvard Univ., fall 1938. Named Prof. of the Practice of Civ. Engr. July, 1946. Consultant, Chicago Subways, 1938-40; Necaxa Dam, Tampico, Mexico 1942; paper mill, Port Alberni, British Columbia, Canada, 1946- ; International Airport, Stockholm, Sweden, 1946; Polavaram Dam, Madras, India, 1946; Brazilian Hydroelectric Authority, 1947-50; Sariyar and Seyham Dam projects, Turkey 1947; Serre Poncon Dam, France, 1948; Sasumua Dam, Kenya, 1953-56; chair, Bd. of Consultants, High Aswan Dam, Egypt, 1954-59. Awards: Hon. mem., ASCE, Boston Soc. of Civ. Engrs. and the Austrian, Mexican and Turkish Societies. Hon. pres. of the Intl. Soc. of Soil Mechanics and Foundation Engineering. Hon. doctorates from Trinity College, Dublin, Ireland, Technical Univ. of Istanbul,

Turkey, National Univ. of Mexico, Federal Polytechnical Univ. of Switzerland, Lehigh Univ., Bethlehem, PA, and the Technical Univ. of Berlin. Norman Medal, ASCE (four times); FitzGerald Medal, Boston Soc. of Civ. Engrs. (twice); Thomas Fitch Rowland Prize, ASCE; Gold Medal of Honor, Austrian Society of Mech. Engrs. and Architects. Publ: Theory of Consolidation (1924); Erdbaumechanic auf Bodenphysikalisher Grundlage (1925); "Principles of Soil Mechanics" (in *Engineering News-Record*, 1926); Rock Tunneling with Steel Supports (1946); Theoretical Soil Mechanics (1946); *Soil Mechanics in Engineering Practice* (with Ralph Peck, 1948). Refs: ASCE files.

THOMPSON, SANFORD ELEAZER; b. Ogdensbury, NY, Feb. 13, 1867; d. _____, Feb. 25, 1949; f. _____; m. _____; w. _____; c. _____. He received a B.S. in civ. engr., MIT in 1889. Immediately went to work on supervision of construction of a paper mill in ME where he was introduced to industrial engineering and scientific management by Frederick W. Taylor (so-called "Father of Scientific Management"). Co-authored *Concrete, Plain and Reinforced* with Taylor in 1905, *Concrete Costs* in 1912. Maj. factor in the formation of Committee C-9 in 1914 and was its 1st chair. He received honors both domestic and foreign from many organizations and societies. Appointed by Herbert Hoover to "Committee on Elimination of Waste in Industry" to study methods of stabilizing production and distribution. Pres. Harding appointed him to the "Economic Advisory Board" and the U.S. Coal Commission had him investigate underground management of bituminous coal mines. He retired from active business only a short time before his death. Refs: HN.

THOMSON, REGINALD HEBER; b. Hanover, IN, March 20, 1856; d. Seattle, WA, Jan.7, 1949; f. Samuel; m. Magdeline Sophronia Clifton; w. Adeline Laughlin; c. James, Marion, Reginald Heber Jr., Frances. Attended Hanover College. Grad. with B.A. 1877; M.A. (A.M.) 1903; Ph.D. (hon.) 1903. Did surveying in CA and taught at Healdsburg Inst. in 1877. Came to Northwest and worked in surveying as Deputy U.S. Mining Engineer for coal in 1881. Asst. city engr./surveyer, Seattle, WA, 1882-84. City surveyor, Seattle (title changed in 1890 to City Engineer), 1884-1886. Engineer for Seattle, Lake Shore and Eastern RR (to become Snoqualmie Branch of Northern Pacific, 1887. City Engineer of Seattle, 1892-1911. Work included survey of Snoqualmie Falls and Cedar River for hydroelectric power, north trunk sewer plans, rail service plans (Great Northern Franchise negotiation), waterfront plaza, introduction of concrete walks and hard surface pavement for city sts., Cedar River water supply drain and pipelines, Density regrade (5.4 million cubic yds.), south trunk sewer tunnel, Cedar River hydroelectric power and municipal lighting system (1902), Great Northern RR Tunnel, refuse plan including design and const. of refuse destructor plants (Meldrum Incinerator), Jackson and Dearborn regrades (3.3 million cubic yds.), navigation improvements in Duwamish River Valley, Lake Washington ship canal/locks system. 1st engr., Port of Seattle, 1912. Layout of Strathcona Park on Vancouver Island, Canada, 1913. Consultant on Lake Washington Pontoon Bridge and foundation of Tacoma Narrows Bridge, Eugene, OR, hydroelectric plant; Rogue River Irrigation Canal, Prince Rupert, British Columbia, Canada, 1914. Mem. ASCE 1903 (hon. mem. 1940; dir., 1917-19); Canadian Soc. of Civ. Engrs.; Univ. of Washington Bd. of Mgrs. of YMCA, 1905-15; Seattle City Council, 1916-21; Good Roads Assn. (pres., 1910), Pacific

Northwest Soc. of Civ. Engrs. (pres., 1902-1903). Refs: RHT, MP, CE1, CE2, BOS.

TOTTEN, GEORGE MUIRSON; b. New Haven, CT, May 28, 1809; d. NYC, May 17, 1884; f. Gilbert; m. Mary (Rice); w. Harriet Seely; c. two daughters. Attended schools in New Haven and the Hopkins Grammar School. Entered the Norwich Military Academy, 1824. Grad. 1827. Asst. engr. on Farmington Canal, 1827. Asst. engr. on Juniata Canal in PA, 1828-31. Asst. engr. on Delaware & Raritan in NJ, 1831-35. Asst. in const. of road from Reading to Port Clinton, PA, 1835. In similar work in VA and PA in connection with Sunbury & Danville RR, 1837-40. In NC in connection with Gaston & Raleigh RR, 1840-43. Received South American commission to the harbor of Cartagena in Columbia, 1843. Engr. in chief of the Canal del Dique to connect harbor with the Magdalene River. Chief engr. of const. of Panama RR, 1850-55. Consulting engr. until his death. Made surveys for a canal across the isthmus, showed that mean level of the Atlantic and Pacific Oceans was the same, 1857. Went to Venezuela on RR work near Caracas. Appt. chief of US Army Corps of Engrs. to make location of Panama Canal, 1878. After location determined, served as consulting engr. Chief engr. of "Sassafras Route" and the Maryland and Delaware Canal. Presented with a ring from Napoleon III, and a gold medallion from Gen. Blano, pres. of Venezuela. Mem: Am. Philosophical Soc. Refs: DAB, CAB, DR, WAB, WWW.

A SPREADER AT WORK IN 1907 ON THE ISTHMIAN (LATER PANAMA) CANAL COMMISSION RAILROAD

TRAUTWINE, JOHN CRESSON, JR.; b. Philadelphia PA, March 17, 1850; d. July 4, 1924; f. John Cresson; m. Eliza Ritter; w. Lucy Smith; c. John Cresson III. Educated in private schools. Trained in commercial work employed in firm of Morris Wheeler and Co., iron brokers, then assisted father in revising and editing *Civil Engineer's Pocket Book*. Translated Ganguiller and Kutter's *Flow of Water in Channels*. Translated parts of H. Bazin's *Flow of Water Over Wiers*, 1890. Collaborated with Olga and Boris Smith in production of paper on "Water Hammer," 1890. Editor of Journal of the Assn. of Engineering Societies. Chief engr. of Bureau of Water, Philadelphia, 1895-99. Served on civ. service bds. in Philadelphia in examining candidates for municipal engr. positions, 1892-1907. Investigated, reported and advised on engr. projects including water distillation plant at Pittsburgh, PA, location of an electric car line in northeast Philadelphia, condition and utilization of leased lines of Mine Hill and Schuykill Haven RR Co., and public water supplies in Southern NJ, Mount Sterling, KY, Red Bank, NH and Cumberland, MD. Advised respecting utilization of the Croton Water Shed and necessity of developing other sources of supply at request of Gen. Committee on Water Supply of NYC, Nov. 1898. Investigated plans and prospects of Rochester and Lake Ontarion Water Co., Rochester, NY, 1909. Mem: ASCE, Franklin Inst. (life mem., mgr.), Philadelphia Engrs. Club (secy., pres.), Am. Soc. of Mech. Engrs., Am. Water Works Assn. and New England Water Works Assn., Soc. of Civ. Engrs. of London, City History Soc. of Philadelphia, Historical Soc. of Pennsylvania, Philibiblon Club, Economic Club (pres.). Publ: Articles for a variety of publications. Refs: TASCE, WWW.

TRIMBLE, ISSAC RIDGEWAY; b. Culpeper County, VA, May 15, 1802; d. Baltimore, MD, Jan. 2, 1888; f. John; w. Maria Cattell Presstman; w. 2nd Ann Ferguson Presstman; c. two sons. Move to KY with family, 1805. Appt. to US Military Academy, grad. 1822. Commissioned 2nd lt. in Artillery, July 1, 1822. Surveyed road from Washington, DC to Ohio River, 1822. Served on ordnance duty, 1822-23. In garrison at Ft. Lafayette, NY, 1823-24. On topographical duty, 1824-30. At artillery school for practice, Ft. Monroe, NA, 1830-31. Resigned from army to work on RRs, May 31, 1832. Principal asst. engr. on Boston & Providence RR, 1832-35. Chief engr. of Baltimore & Susquehanna RR, 1835-38. Chief engr. of York & Wrightsville RR, PA, 1836-38. Chief engr. of Philadelphia, Wilmington & Baltimore RR, 1842-53. Chief engr. of Philadelphia & Baltimore Central, 1854-59. Gen. supt. of Baltimore & Potomac RR, 1859-61. Involved in RR operations in West Indies. Returned when the Civil War began in 1861. Burnt bridges north of Baltimore to obstruct movement of Union troops. Commissioned as col. of engrs. in state troops, May 1861. Commissioned brig. gen. in Confederate Service, Aug. 9, 1861. Charged with const. of batteries on Potomac River to prevent passage of US vessels, Sept. 1861. Assigned to command a brigade in Confederate Army of northern VA and charged with removal of stores from depot at Manassas Junction when the army was withdrawn, 1862. Took part in operations in the Shenandoah Valley, 1862. In the battles of Winchester and Cross Keys. Took part in the Seven Days' battles. In the campaign against Pope's Army in northern VA, 1862. In the battles of Cedar Mountain, Hazel Run and in Jackson's March around Pope to Bristoe Station on the Orange & Alexandria RR. Captured Union Depot of supplies at Manassas Station, Aug. 27, 1862. Seriously wounded at battle of Manassas. Promoted to command of troops in the Shenandoah Valley, June 18,

1863. Assigned to command division in Hill's Corps when the cmdr. was seriously wounded. Took part in 3rd day of battle at Gettysburg. Seriously wounded, lost a leg and taken prisoner. Exchanged in Feb. 1865. On his way to join Lee when the surrender occurred at Appomatox. Refs: DAB, WAB, WWW, CAB, DR, BDNA.

TSCHEBOTARIOFF, GREGORY P.; b. Pavlovsk, Russia, Feb. 15, 1899; f. Porphyry G.; m. Valentina I. (Doubiagsky); w. Florence D. Bill. Grad. from Imperial Law School of Petrograd, Russia, May 1916. Wartime course of Mihailovsky Artillery School, Petrograd. Russia, May-Dec. 1916. Degree of civ. engr. from Dept. Technische Hochschule Berlin-Charlottenburg, Germany, Oct. 1921-Dec. 1925. 2nd lt., Don Cossack Battery & Guard Horse Artillery, Dec. 1916-17. Interpreter to Alaman of Don Cossacks, during negotiations with Allied Missions in South Russia, 1918. Officer 2nd Don Cossack Horse Battery of White Armies, Feb.-Sept. 1919. A.D.C. to inspector of artillery, Don Army, Sept. 1919-March 1920. Lt. Adjutant and English teacher, Don Cadet Corps, left Russia for Ismailia, Egypt, March 1920-Sept. 1921. Assisted by British Red Cross and Am. YMCA. Student, Technische Hochschule, Berlin, Germany, Oct. 1921-Dec. 1925. Calculator and draftsman of reinforced concrete for Pelnard-Considere & A. Caquot, consulting engrs., Paris, France, June-Oct. 1924. Engr. of stress computations of steel bridges in Berlin, Germany, March-Sept. 1926. Engr., designer of reinforced concrete structures with Paul Kossel & Co., building contractors, Bremen, Germany, Sept. 1926-Aug. 1927. Engr., designer of reinforced concrete structures with Austrian Building Co. of Cairo, Egypt, Sept. 1927-March 1929. Engr. in charge of shift and shoring of excavations, Habermann & Gukes-Liebold AG on site of const. of lock on Weser River at Hameln, Germany, April-Nov. 1929. Engr. of reinforced concrete and foundation work at the Technical Office, State Buildings, Dept. of Ministry of Public works, Egyptian govt., Nov. 1929-Nov. 1933. Research engr., Foundation Soils Research Laboratory, Facility of Engr., Egyptian Univ., Dec. 1933-Jan. 1937. Official delegate of the Egyptian Univ., presented paper on soil research to 1st Intl. Conference on Soil Mechanics & Foundation Engr., Harvard Univ., June 1936. Lecturer on foundations, Princeton Univ., Feb.-June 1937. Asst. prof. civ. engr., Princeton Univ., June 1937- . Mem: ASCE, Am Soc. for Testing Materials, Am. Concrete Inst., Sigma Xi, Soc. for the Promotion of Engr. Education, Princeton Engr. Assn. Publ: several papers in technical publications. Refs: WWE, Bray.

TURNER, CLAUDE ALLEN PORTER ("CAP"); b. Lincoln, RI, July 4, 1869; d. _____ 1955; f. J. M. Turner; m. Elizabeth (Darling) Turner; w. Mary E. Burns, June 6, 1894; w. 2nd Kathleen Flavin, 1943; c. 1 son, 1 daughter. Grad. B.S. in engr., Lehigh Univ., Bethlehem, PA, 1890. Worked in engr. dept. of New York New England RR 1890. Successively, draftsman with Edgemore Bridge Co., Philadelphia, PA, asst. engr. of Columbus Bridge Co., Columbus, OH, Pittsburgh Bridge Co. (1-2 years), Berlin Bridge Co., Berlin, CT; asst. engr. and draftsman, Pottsville Iron & Steel Co., Pottsville, PA. Moved to Minneapolis about 1897; with Gillette Herzog Co., 1897 until formation of American Bridge Co., 1910; then asst. engr. in charge of design dept. with that firm and engr. of western contracting dept. 1st year of Am. Bridge, 1900. Founder and pres. of Turner Construction Co. in 1902, builder of bridges, sugar mills, reinforced concrete structures of all kinds, manufacturing plants. Patented flat-slab construction (mushroom cap on columns), 1905; granted 30+ patents for reinforcement and reinforced concrete. Member ASCE, Engrs. Club of St. Paul, Engrs. Club of Minneapolis; Bd. of Dirs., Mercantile State Bank. Publ: Concrete Steel Construction (1909), Elasticity: Structure & Strength of Materials (1922-1934). Refs: HN.

TWEEDDALE, WILLIAM; b. Beith, Ayrshire, Scotland, May 18, 1823; f. Edward; m. Janet (Kerr); w. Ellen W. Parker; c. three. Came to NYC with parents, March, 1833. Lived in Albany and Saratoga County. Returned to NYC, Aug. 1835. Engaged in farming until winter 1848. Entered Troy Conference Academy, Poultney, NY. Entered RPI, fall of 1849 for one term, reentered, fall 1851, grad. with degree of civ. engr. in 1853. Asst. in office of city engr., Troy, NY and on const. of Albany and Susquehanna RR until Sept. 1854. Returned to RPI and took charge of field work. Bridge engr. and contractor in Chicago, 1855. Designed and built superstructure of iron roadway swing bridge and wooden RR swing bridge. Moved to Dubuque, IA, Jan. 19, 1860. Const. bridges and buildings on Dubuque & Sioux City RR. Designed and erected Dubuque elevator, fall 1860. Raised co. for engr. regiment at start of Civil War, mustered in as capt. Engr. operations against New Madrid, resulted in its capture. Cut passage for fleet of transports across lower end of Island No. 8., used for transportation of troops across river from New Madrid to operate against Island Co. 10, resulted in its evacuation. In command of advanced parties of engrs. with division in the siege of Corinth. Engaged in engr. work, reconst. of RRs, dredging of rivers, removal of debris at various points across the Mississippi. Promoted to brevet col. of Volunteers, March 13, 1865. Resigned May 31, 1865. Supt. const. of iron bridge over Connecticut River at Warehouse Point. Went back west, Fall 1866. Moved to Chicago, Aug. 1867. Moved to Topeka, KS. Supt. erection of east wing of the state capitol, engaged on same building the following year, 1867-68. Engr. of bridge 900 ft. in length across Kansas River at Topeka. Gen. engr. work as city engr., 1870-78. Engaged in work for west wing of capitol and on foundation for govt. buildings, 1879-80. Employed on same buildings, 1881-83. Refs: RPI Bio, WAB, CAB.

VOSE, GEORGE LEONARD; b. Augusta, ME, April 19, 1831; d. Brunswick, ME, March 30, 1910; f. Richard Hampton Vose; m. Harriet Green (Chandler); w. Abba Valentine Thompson; w. 2nd Charlotte Buxton Andrews; c. four daughters and one son. Educated in high schools at Augusta, ME and Salem, MA. Entered office of Samuel Nott, civ. engr. in Boston. Studied in Lawrence Scientific School at Harvard, 1849-50. Asst. engr. on Kennebec and Portland RRs. Engaged on RR location and const. across the US, 1850-59. Assoc. editor of *American Railway Times*, 1859-63. Lived in Salem, MA for three years. Moved to Paris, ME, 1866. Prof. of civ. engr. at Bowdoin College, Brunswick, ME, 1872-81. Prof. of civ. engr. at MIT, 1881-86. Mem: Boston Soc. of Civ. Engrs. (pres., 1884-87). Publ: *Handbook of Railroad Construction* (1857; one of the earliest on the subject), *Manual for Railroad Engineers and Engineering Students* (1873), *Orographic Geology* (1866), *A Graphic Method for Solving Certain Algebraic Problems* (1875), *An Elementary Course of Geometrical Drawing* (1878), *A Sketch in the Life and Works of Loammi Baldwin* (1885), *Bridge Disasters in America* (1887). Refs: DAB, CAB.

WADDELL, JOHN ALEXANDER LOW; b. Port Hope, Ontario, Canada, Jan. 15, 1854; d. NYC, March 3, 1938; f. Robert Needham; m. Angeline Esther (Jones); w. Ada Everett; c. Needham Everett, Leonard, and Ethel. Moved with family to Cobourg, Ontario, educated by private tutors. Traveled to China and back aboard a clipper for health reasons the year before he entered college. Attended Trinity College School in Port Hope and business college for five months in Toronto. Entered RPI, grad. with degree of civ. engr. in 1875.

JOHN ALEXANDER LOW WADDELL (1854-1938)
INTERNATIONALLY EMINENT BRIDGE DESIGNER

Draftsman in marine dept. of Canadian govt., 1875. Rodman on Canadian Pacific RR, May 1876-July 1877. Contracting engr. on Canadian Pacific, July-Nov. 1877. Asst. prof. of rational and tech. mechanics, RPI in fall 1878-fall 1880. Chief engr. of Raymond & Campbell, bridge builders, Council Bluffs, IA, Jan. 1881-May 1882. Prof. of civ. engr. at Imperial Univ. of Tokyo, 1882-86. Assoc. with Phoenix Bridge Co. in Phoenixville, PA when he returned to the US, 1886-92. Opened office as bridge designer and consultant in Kansas City, MO, 1886-99. Designed or supervised Red Rock cantilever bridge for Atlantic & Pacific RR across Colorado River between AZ and CA, 1889, a double track RR bridge across Missouri River at East Omaha, NB with two 520-ft. swing spans, 1893, Grand Island highway bridges across Niagara River near Buffalo, NY. Consulting engr. on building an elevated RR system in Chicago, 1893. Invented and successfully introduced large-scale high clearance vertical lift bridge, the 1st being the South Halsted St. Bridge in Chicago, 1893. In partnership with Ira G. Hedrick, 1899-1907. Investigated suitability of nickel steel for bridge building for the International Nickel Co., 1903. Partnership with John Lyle Harrington,

1907-15, and his son John Lyle Harrington, 1915-18. Moved office to NY, 1920. Practiced alone, 1920-27. Member of jury of awards in competition for new Yellow River Bridge for Peking-Hankow RR, 1921. Parnership with Shortridge Hardesty, 1927-38. Designed and supervised const. of Mississippi Highway Bridge across the Mississippi River, 1929 and Anthony Wayne High Level Bridge at Toledo, OH, a suspension span across Maumee River, 1931. Consulting engr. and technical advisor to Ministry of RRs, National Government of China, Nanking, 1929. Hon. technical advisor to China, 1929-38. Consulting engr. on Outerbridge Crossing and Goethals Bridge, connecting Staten Island with NJ, 1928, and Marine Parkway Bridge across Rockaway Inlet, 1936-37. Designed bridges in Canada, Mexico, Russia, China, Japan and New Zealand. Hon. degrees of B.A. and M.Engr. in 1882 and D.Sc. 1904 from McGill Univ.; LL.D. in 1904 from Univ. of Missouri; D.Engr. in 1911 from the Univ. of Nebraska; D.Engr. in 1915 from Imperial Univ. of Japan; D.Litt. in 1934 from Univ. of Puerto Rico. Knight Cmdr., Order of the Rising Sun, Japan, 1888. Knight 1st Class, Order of Societe de Biengaisance of Grand Duchess Olga of Russia for services as principal engr. of Trans-Alaskan-Siberian RR Project, 1909. 2nd Class, Order of Sacred Treasure of Japan, 1921. 2nd Class, Order of Chia Ho, China, 1922. Cavaliere of the Crown of Italy, 1923. Clausen Gold Medal of Am. Assn. of Engrs., 1931. Mem: ASCE, Am. Railway Engr. Assn., Engr. Inst. of Canada, Am. Inst. of Consulting Engrs., Am. Soc. for Testing Materials, Intl. Soc. for Testing Materials, Soc. for Promotion of Engr. Education, Inst. of France, Soc. of Civ. Engrs. of France, Chinese Inst. of Engrs., Inst. of Structural Engrs. of Great Britain, National Engr. Soc. of Barcelona, National Engr. Soc. of Peru, Kogaku Kyokai of Japan, Rensselaer Soc. of Engrs., Western Soc. of Engrs., Conservation Assn., National Economic League, Tau Beta Pi, Sigma Xi, Phi Beta Kappa, Phi Tau Phi, Engrs. Club of Kansas City, MO, and Authors Club of London, England. Publ: *The Designing of Ordinary Iron Highway Bridges* (1884), *A System of Iron Railroad Bridges for Japan* (1885), *De Poutibus* (1898), *Specifications for Steel Bridges* (1900), *Engineering Specifications and Contracts* (1907), "Nickel Steel for Bridges" (Norman Medal, ASCE, 1909), "Economics of Steel Arch Bridges" (Norman Medal, ASCE, 1918), *Bridge Engineering* (1916), *Economics of Bridgework* (1921), *The Principal Professional Papers of Dr. J. A. L. Waddell* (1905), *Memoirs and Addresses of Two Decades* (1929). Refs: DAB, WWW, WAB, RPI Bio, WWE.

WAITE, CHRISTOPHER CHAMPLIN; b. Maumee City, OH, Sept. 24, 1843; d. Feb. 21, 1896; f. Morrison Remick; m. Amelia (Champlin); W. Lillie C. Guthrie; c. Harry and Ellison. Entered RPI in 1860, grad. in 1864. Asst. engr. on Rensselaer & Saratoga RR, July 1864. Asst. engr. on Croton Water Works Bd., 1865. On the Coxsacie RR, 1865-66. Chief engr. of Columbus & Toledo RR, 1867. Chief engr. of Toledo, Akron & Atlantic RR, 1868. Cashier of P. C. & St. Louis RR, 1869. Chief engr. of the Cincinnati & Muskingham Valley RR, 1870. Supt. of the Cincinnati & Muskingham Valley RR and Little Miami RR, 1881. Asst. to pres. of NY, L.E. & W. RR. Vice pres. of Cincinnati Chamber of Commerce. Trustee of the Children's Hospital. Pres. of Bd. of Trustees of OH State Epileptic Hospital. Mem: ASCE. Refs: RPI Bio, TASCE.

WALKER, REUBEN LINDSAY; b. Logan, VA, May 29, 1827; d. VA, June 7, 1890; f. Meriwether Lewis; m. Maria (Lindsay); w. Maria Eskridge; w. 2nd Sally Elam; c. eight. Grad. from the Virginia Military Inst., 1845. Engr. on Chesapeake and Ohio RR, 1845-61. At start of the Civil War, capt. of Purcell Battery and sent to Aqui Creek, VA. Fought at Bull Run, July 21, 1861. With his battery in VA for the rest of 1861. Promoted to maj., March 1862. Chief of artillery for A. P. Hill's division in 2nd battle of Manassas, Aug. 29-30, 1862. At capture of Harper's Ferry, Sept. 14-15, 1862. Fought at Fredericksburg. Promoted col. Became chief of artillery. Commanded reserve artillery of Hill's 3rd Corps at Gettysburg, July 1-3, 1863. In remaining campaigns in VA. Surrendered with Lee at Appomotox. Appt. brig. gen. of artillery, Feb. 1865. Engaged in farming after the war. Moved to Selma, AL, 1872. Supt. of Marine & Selma RR, 1872-74. Returned to VA in 1876. Employed by Richmond & Danville RR, 1876-77. Supt. of Richmond St. RRs. Const. engr. for Richmond & Allegheny RR. Supt. building of women's dept. of the VA State Penitentiary. Appt. supt. of const. of the TX State Capitol, lived in Austin until 1888. Went back to farm in VA. Refs: DAB, WWW, Lamb, BDNA, CAB.

WARREN, GOUVENEUR KEMBLE; b. Cold Spring, NY, June 8, 1830; d. Newport, RI, Aug. 8, 1882; f. Sylvanus Warren; w. Emily Forbes Chase; c. two. Educated in Cold Spring and across Hudson at Kinsley's School. Appt. to US Military Academy in 1846, grad. second in his class, July 1, 1850. Appt. brevet 2nd lt. in Corps of Topographical Engrs. Asst. engr. on survey of the Delta of the Mississippi River, 1850-54. Mem. of bd. of improvement of the canal around Falls of the Ohio. Head of surveys for improvement of Rock Island and Des Moines Rapids. Compiler of maps and reports of Pacific RR Exploration. Promoted to 2nd lt., Sept. 1, 1854. Chief topographical engr. of Sioux Expedition, 1855. Battle of Blue Water, Sept. 3. Promoted 1st lt., July 1, 1856. Made maps and reconnaissances of Dakota Territory and Nebraska Territory until Aug. 1859. Asst. prof. of mathematics at Military Academy, Aug.-Nov. 1859, principal asst. prof, 1859-61. Lt. col. of 5th New York Volunteers, May 14, 1861. In action at Big Bethel Church, June 10. Const. of defenses around Baltimore and Washington. Col. of his regiment, Aug. 31, 1861. Capt of topographical engrs., US Army, Sept. 9, 1861. Engaged in siege of Yorktown, April-May 1862. Commanded a brigade at Pamunky River and Hanover Court House, May 26-27, 1862. Wounded at Gaines' Mill, June 27. Brevetted lt col., US Army. Commanded repulsion of Wise's Division at Malvern Hill. In engagement at Harrison's Landing. Took part in 2nd Battle of Bull Run and Centervillle, Aug. 30-Sept. 1, 1862. Commanded brigade in Maryland Campaign, Sept.-Nov. 1862. Brig. gen. of Volunteers, Sept. 26. Served at battle of Fredericksburg in Dec. Chief topographical engr. of Army of the Potomac, Feb. 4, 1863. In battle at Orange Pike, Marye Heights and Salem in May. Promoted maj. gen. of Volunteers, June 3, 1863. Chief engr., Army of the Potomac, June 8-Aug. 12, 1863. At Gettysburg, July 1-3, 1863. Brevetted col., US Army for services at Gettysburg. Wounded in defense of Little Round Top. In command of the V Corps, March 24, 1864. In action of the Wilderness, Spotsylvania, Cold Harbor and the assaults on Petersburg. Promoted maj., US Army, June 25, 1864. Maj. gen., US Army, March 13, 1865. Relieved of command at Five Forks, April 1, 1865. Transferred to command defenses of Petersburg and the Southside RR, April-May, then commanded Dept. of the Mississippi, May 14-30, 1865. Resigned volunteer commission and reverted to maj. of engrs., US Army. Prepared maps and reports of campaigns and wrote for publication results of some of his early explorations. Mem. of bd. of engrs. to examine canal at Washington, DC as supt. engr. of surveys and improvements of upper Mississippi. Mem. of commission to examine Union Pacific RR and Telegraph Lines. In charge of survey of the battlefields of Gettysburg. Supervised building of Rock Island Bridge across the Mississippi, 1869-70. River and harbor work for Corps of Engrs. in the upper Mississippi Valley, along Atlantic Coast and in the Great Lakes. Mem. of the advisory council of Harbor Commission of RI, Oct. 10, 1878. Promoted lt. col. of engrs., March 4, 1879. Request for inquiry given Dec. 9, 1879, vindicated Nov. 21, 1881, three months after his death for relief of his command at Five Forks by Gen. Sheridan. Statue of Warren unveiled at Gettysburg, Aug. 8, 1888. Mem: ASCE, Am. Assn. for the Advancement of Science, Am. Philosophical Soc., and the National Academy of Sciences. Publ: "Examination of Reports of Various Routes" in *Reports of Explorations and Surveys for a Railroad to the Pacific Ocean* (1855), *Memoir to Accompany the Map of the Territory of the United States from the Mississippi River to the Pacific Ocean, Giving a Brief Account of Each of the Exploring Expeditions Since A.D. 1800* (1859), *An Account of the Operations of the Fifth Army Corps* (1866), *Report of the Survey of the Upper Mississippi River and its Tributaries* (1867), *An Essay Concerning Important Physical Features Exhibited in the Valley of the Minnesota River* (1874), *Preliminary Report of Explorations in Nebraska and Dakota in the Years 1855-56-57,* (1875), *Report on the Transportation Route Along the Wisconsin and Fox Rivers Between the Mississippi River and Lake Michigan* (1876), *Report on Bridging the Mississippi River Between St. Paul, Minnesota and St. Louis, Missouri* 1(878). Refs: Lamb, WWW, DAB, WAB, CAB, Lamb, DR, BDNA.

WATKINS, JOHN ELFRETH; b. Ben Lomond, VA, May 17, 1852; d. Aug. 11, 1903; f. Francis B; m. Mary (Elfreth); w. Helen Bryan; w. 2nd Margaret Virginia Gwynn; c. five. Educated at Tremount Seminary, Norristown, PA. Grad. from Lafayette College with degree of civ. engr. in 1871. Mining engr. of the Delaware & Hudson Canal Co., 1871-72. Asst. engr. of const. with Pennsylvania RR Co., 1872-74. Lost right leg in an accident, disabled 1873. Degree of M.S. from Lafayette College, 1874. Amboy Division of the Pennsylvania RR. Chief clerk of Camden & Atlantic RR. Reassigned to Amboy Division of Pennsylvania RR until 1886. Assoc. with US National Museum at Washington, DC, hon. curator of transportation, 1884. Salaried position, 1886-92. Promoted US Patent Office Centennial Celebration, 1891. Prepared Pennsylvania RR Co. exhibit for World's Columbian Exposition, Chicago, 1893. In charge of dept. of industrial arts in Field Museum, Chicago. Returned to Washington, DC as curator of mech. technology and supt. of buildings of National Museum, 1895-1903. Wrote history of Pennsylvania RR, 1845-96, never published, uncompleted when he died. Publ: "The Beginnings of Engineering" (1891), *The Development of the American Rail and Track as Illustrated by the Collection in the US National Museum* (1891), *Log of the Sacannah Museum* (1891). Refs: DAB, CAB, WWW.

WEBB, GEORGE HERBERT; b. Dubuque, IA, March 5, 1860; d. near Boston, MA, Nov. 3, 1921; f. George Webb; m. Emma (Alder); w. Jessie Lawrence. Moved to Johnstown, PA with family, 1876. Entered Pennsylvania Military College in Sept. 1877, grad. with degree of civ. engr., June 10, 1880, M. civ. engr., 1914. Rodman on Baltimore & Ohio RR Co. Tested steel for Brooklyn Bridge, 1881. Levelman and transitman Pittsburgh Southern RR and Pittsburgh & Western RR. City engr., Johnstown, PA, 1883. Asst. engr. of Cambria Iron and Steel Co. Transitman, locating engr. and division engr. on const. of Chicago, Burlington & Quincy RR Co., fall 1885-spring 1888. Division engr. in charge of const. of sections of Seattle, Lake Shore & Eastern RR, 1888. Const. of govt. RRs in Chile for North and South America Const. Co., spring 1889-91. Division engr. and supt. of const. for Summit Division of Central RR of Peru through main range of the Andes, 1891-93. In private practice, 1893-97. Chief engr. of Cincinnati, Georgetown & Portsmouth RR, 1897-99. Roadmaster of Cleveland, Cincinnati, Chicago & St. Louis and Chicago & Alton RRs, 1899-1900. Const. engr. of Baring Cross shops of St. Louis, Iron Mountain & Southern RR, AR, 1901-1902. Location of Ostimo Division and engr. of Middle Division of Michigan Central RR Jan.-Nov. 1903, asst. chief engr., 1903-1905, chief engr., June 1905-17, 1919-21. Chief engr. of Detroit River Tunnel Co., Jan. 16, 1911. Lt. col. of engr. Corps of US Army, July 6, 1917. Assigned to 16th RR Engrs. Col. Sept. 27, 1918. Section engr. with A.E.F. in France, Aug. 1917-April 1919. Hon. M.S. in civ. engr., Pennsylvania Military College, June 17, 1914. Awarded Distinguished Service Medal and the French Ordre de l'Ecole Noire (Officier). Mem: ASCE, Am. Railway Engr. Assn. (dir.), Detroit Chapter of the Military Order of the World War, 16th Engrs. Post of the Veterans of Foreign Wars of the US, and the Am. Legion. Refs: WWW, TASCE.

WEBSTER, GEORGE SMEDLEY; b. Philadelphia PA, Oct. 19, 1855; d. Philadelphia, Jan. 23, 1931; f. John Hambleton; m. Lydia (Smedley); w. Mary Heston Anderson; c. Maurice Anderson. Early education at Friends Select School in Philadelphia. B.S. from Univ. of Pennsylvania, 1875. Member of engr. corps of the Centennial Exposition, Philadelphia, 1875-76. Asst. engr. of US Coast and Geodetic Survey, 1876-77. Asst. engr. of Philadelphia, 1877-92, principal asst. engr. and acting chief engr., 1892-93, chief engr. and pres. of bd. of surveyors, 1893-1916, 1920-21. Designed and const. a number of large bridges, including concrete bridge over Wissahickon Creek and Gray's Ferry and Passyunk Ave. bridges over the Schuykill River. Studied sewage treatment plants in Europe for one year. Built 1st plant in Philadelphia, 1908. Acting chief of Bureau of Filtration. Dir. of Dept. of Wharfs, Docks and Ferries, Philadelphia, 1916-20. Mem. bd. of engrs., Delaware River Bridge Joint Commission, Jan. 1, 1921. Mem., PA State Sanitary Water, ND, 1927. Pres. of bd. of mgrs., Friends Hospital. Technical vice pres., Regional Planning Federation of Philadelphia, Tri-State District. Hon. D.Sc. from the Univ. of Pennsylvania, 1910. Mem: ASCE (pres., 1920-21), Am. Soc. for Testing Materials (pres., 1920-21), Franklin Inst., Engr. Alumni Assn. Univ. of Pennsylvania (pres., 1914), Philadelphia Engrs. Club (pres., 1895), Municipal Engrs. Soc. (pres., 1914), Sanitary Engr. Section of the Am. Public Health Assn. (pres., 1918), Division of Engr. of the National Research Council, National Conference on City Planning, Fairmount Park Art Assn., Pennsylvania Forestry Assn., Union League Club of Philadelphia. Publ: "Pennsylvania Avenue Subway and Tunnel" (1898), "Development of the Delaware River Water Front" (1902),

"The Walnut Lane Bridge" (1909), "Report on the Treatment of the Sewage of the City of Philadelphia" (1914), and other papers. Refs: WWW, WWE, WAB.

WEITZEL, GODFREY; b. Cincinnati, OH, Nov. 1, 1835; d. Philadelphia, PA, March 19, 1884; f. Louis; m. Susan; w. 2nd Louisa Bogen; c. one daughter. Educated in local schools. Entered US Military Academy in 1851, grad. 2nd in his class on July 1, 1855. Commissioned 2nd lt., July 27, 1856. 1st lt., July 1, 1860. 1st duty on fortifications of New Orleans, 1855-59. Asst. prof. of engr. at US Military Academy, 1859-Jan. 1861. Engr. duty in Washington, 1861. In expedition to Pensacola, FL, April 19-Sept. 17, 1861. Chief engr. of fortifications of Cincinnati, 1861. Returned to Washington, DC in command of an engr. co. Chief engr. of Gen. Butler's force, spring 1862. Asst. military cmdr. of New Orleans. Brig. gen. of Volunteers, Aug. 29, 1862. Engaged in field operations until Dec. 1863. Commanded a brigade and provisional division in siege of Port Hudson. In assaults of May 27 and June 14, 1863. Captain in regular engr. corps. Brevet maj. and lt. col. for Thibodeaux and Port Hudson. In command of 2nd Division XVIII Army Corps in Butler's Army of the James, May-Sept. 1864. Chief engr. of that army. Supervised const. of defenses of Bermuda Hundred. Brevet maj. gen. of Volunteers, Aug. 1864. Returned to troop duty, Sept. 1864. Commanded XVIII and later XXV Army Corps. Brevet rank of col. in regular service, Sept. 29, 1864. Promoted maj. gen. of Volunteers, Nov. 17, 1864. Second in command to Butler in 1st expedition against Ft. Fisher. Took possession of Richmond on evacuation, April 3, 1865. Brevetted brig. gen. and maj. gen. in regular army. Commanded Rio Grande District. Mustered out of volunteer service, March 1, 1866. Returned to Corps of Engrs. Maj., Aug. 8, 1866. Engaged in river and harbor improvement. Connected with ship canals at falls of the Ohio and at Sault Ste. Marie, MI, and lighthouse at Stannard's Rock in Lake Superior, 1867-73. Lt. col., June 23, 1882. Transferred to Philadelphia. Publ: translations of German works dealing with hydraulic engr. and canal const. Refs: DAB, WWW, DR, WAB, Lamb, CAB.

WESTON, CHARLES VALENTINE; b. Kalamazoo, MI, Feb. 14, 1857; d. Jan. 27, 1933; f. John; m. Catherine (Clark); w. Catherine Dyer; w. 2nd Olga Thimm; c. Charles Edward, Florence Elizabeth. Educated in public schools of Kalamazoo. Rodman with Sabine Pass & Northwestern RR, Texas, 1879-80. Asst. engr. on Missouri-Kansas-Texas RR, 1880-81. Asst. engr. on Kansas City, Springfield & Memphis RR, 1881-82. Asst. engr. for Chicago & North Western RR, 1882-84. Asst. engr. for Kansas City, Clinton & Springfield RR, 1884-86. Division engr. in charge of const. of Gulf, Colorado & Santa Fe RR in TX, 1886-88. In charge of const. of intake crib and water tunnel under Lake Michigan for municipality of Lake View, 1888-90. Const. of West Chicago St. RR tunnel under the Chicago River, 1890-94. Chief engr. of Northwestern, Lake St. and Union elevated lines, 1894-1901. Began firm with his brother George, Weston Bros., consulting and const. engrs., 1901-1903. Chief engr. of the South Side elevated RR in Chicago, 1903-1908. Pres. and gen. mgr. of the co., 1908-11. Representative of the bd. of supervising engrs. in charge of traction development, 1907. In private practice, 1911. Operating mgr. of Market St. elevated RR in Philadelphia, 1918-20. Consulting engr. of Chicago Surface lines and a dir. of Chicago RRs Co., 1920-33. Mem: ASCE, Am. Inst. of Mining and Metallurgical Engrs., Western Soc. of Engrs., Masonic Order and Union League Club of Chicago. Refs: WAB, WWE, WWW.

WHITNEY, CHARLES SMITH, b. Bradford, PA, Nov. 4, 1892; d. Paris, France, Oct. 25, 1959; f. Henry Parker Whitney; m. Myra Eva Allen; w. Gertrude Schuyler; c. James Schuyler, Lilian Randolph, and Charles Allen. Graduated Cornell Univ. with B.S. in civ. engr., 1914; M.S. in civ. engr., 1915. As a student, worked for the city engrs. in Bradford, PA, 1910-13; for Stone & Webster as a concrete detailer on MIT buildings, 1914; and for Gustav Lindenthal, 1915, as an inspector of the Hell Gate Bridge and the Bronx Viaduct for the NY Connecting RR, where he met Othmar Ammann. Asst. engr. in Los Angeles, CA for architect John Parkinson working on the steel and concrete design of the Union Terminal buildings, 1916-17. He joined the U.S. Army and served in World War I with the Corps of Engrs. constructing light RRs in France, 1917-19. Moved to Milwaukee, WI, in 1919 to be chief engineer for architectural firm A. C. Eschweiler. Became chief engr. and mgr., Hool & Johnson, Mech., 1920-22. Est. self in structural engr. practice in Milwaukee, WI, 1922. As const. engr. for the city, he became a specialist in municipal planning and design. Projects included the Milwaukee County Expressway system, Lincoln Memorial Bridge, Milwaukee Memorial Center, an athletic bldg. for the Univ. of Wisconsin at Madison, as well as bridges, sewers, water systems, and municipal improvements. Developed expertise in the theory and practice of reinforced concrete design, including shells, domes, and arches, using ultimate strength design and plastic theory. Early projects include Lakeside Park Bridge in Fond Du Lac, Range Line Road Bridge in River Hills, Highland Ave. Bridge in Ozaukee, Milwaukee County Expressway system, and the "Art Nouveau" Sixth St. Bridge in Racine. In 1941, chief architect, planning, for Camp McCoy, WI, a $30 million project, in partnership with Mead, Ward & Hunt Architects-Mech. After 1946, in partnership with Othmar Ammann in NYC, retaining his office in Milwaukee, est. Ammann & Whitney, engineered the Onondaga Memorial Auditorium, Syracuse, NY; Alabama State Coliseum in Montgomery; Athens Airport; both Dulles Airport and the TransWorld Airlines Terminal at John F. Kennedy Airport for architect Eero Saarinen, 1952; and a series of unique, cantilevered airplane hangars which provide large, unobstructed floor areas, for TransWorld Airlines, Pan Am, and others airlines. In 1952, led firm overseas, starting with the US Air Force bases program in France. Pioneered ultimate strength design, arch, shell, folded plate and cable-suspended long span roof structures of reinforced concrete. Awards: ASCE's James R. Cross Medal (1925), Cornell Univ.'s Fuertes Graduate Medal (1925 and 1927), Am. Concrete Inst. Alfred Lindau Award (1951) and Wason Medal (1932, 1952, and 1955), Concrete Reinforcing Steel Inst. 1st Annual Award (1949), Am. Inst. of Architects Allied Professional Gold Medal, awarded posthumously in 1962. Registered professional engr., WI, OH, NY; registered civ. engr., WI; registered structural engr., IL. Mem: ASCE, (pres., Milwaukee Section, 1930) and chair, ASCE Committee on Thin Shell Design; Chair, joint ASCE-ACI Committee on Shear and Diagonal Tension; Am. Concrete Inst., (dir., vice pres., pres., 1955), Chair, Am. Concrete Inst. Committee on Plain and Reinforced Concrete Arches; Am. Soc. for Testing Materials, Soc. of Prof. Engrs., Wisconsin Soc. of Prof. Engrs., Am. Assn. of Airport Executives, Intl. Assn. for Bridge and Structural Engrg., Soc. for Experimental Stress Analysis, Inst. of Aeronautical Sciences; Am. Road Builders Assn., Am. Rway Engr. Assn., Am. Shore and Beach Preservation Assn., Tau Beta Pi, Sigma Xi, Highway Research Bd., National Research Council, Wisconsin Chapter, Am. Inst. of Architects (hon. mem.). Publ: *Bridges,*

A Study in their Art, Science, and Evolution (1928), *Design of Symmetrical Concrete Arches* (1925), "Design of Reinforced Concrete Members Under Flexure or Combined Flexure and Direct Compression" (1927, ACI, Vol. 33), Cost of Long-Span Concrete Shell Roofs (1950), Comprehensive Numerical Method for the Analysis of Earthquake-Resistant Structures (1951), "Reinforced Concrete Thin-Shell Structures" (ACI, Vol. 51), "Guide for Ultimate Strength Design of Reinforced Concrete" (ACI, Vol. 53), "Ultimate Shear Strength of Reinforced Concrete Flat Slabs, Footings, Beams, and Frame Members Without Shear Reinforcement" (ACI, Vol. 54), "Cantilevered Folded Plate Roofs" (ACI, Vol. 55).

WILLIAMS, GARDNER STEWART; b. Saginaw, MI, Oct. 22, 1866; d. Ann Arbor, MI, Dec. 12, 1931; f. Stewart Beach; m. Juliet Merritt (Ripley); w. Jessie Benton; c. Harriet Ripley, William Wright. Grad. from Saginaw High School, 1884. Attended Univ. of Michigan, grad. 1889. Degree of civ. engr., 1899. Asst. engr. of water works const. in Bismarck, ND, 1887. Engr. at Russel Wheel & Foundry Co., Detroit, MI, 1889-93. Civ. engr. to Bd. of Water Commissioners of Detroit, MI, 1893-98. Found source of serious typhoid epidemic of 1892. Experimented with flow of water in pipes and in original dam designs. Prof. of experimental hydraulics and in charge of hydraulic laboratory at Cornell Univ., 1898-1904. Designed and built one of the 1st mech. water filtration plants in the US, and the 1st dome dam at Six Mile Creek, Ithaca, NY, 1902. Mem. of Intl. Waterways Commission, 1903-1905, made a study of flow of water over dams. Expert in Drainage Canal Case of Missouri vs. Illinois and the Sanitary District of Chicago, 1903. Prof. of civ., hydraulic, and sanitary engr. at the Univ. of Michigan and head of the dept. of civ. engr., 1904-11. Expert in the US Circuit Court of IL in the case of US vs the Sanitary District of Chicago, 1907-14, and in hearings before the Supreme Court, 1927. Introduced vertical water wheel in an open flume setting, directly connecting to an electric generator at Sault Ste. Marie, MI. Private consultant with headquarters at Ann Arbor, 1911-31. Built multiple arch dams at Sturgis, MI, in 1910 and on the Ural River in the Soviet Union, 1930. Commissioned maj. in Engr. Officers Reserve Corps, US Army, 1917. Served on Committee on Industrial Preparedness under Council of National Defense. Chief engr. on const. of Camp Beauregard, Alexandria, LA, 1917. In Am. Engr. Council, chair of Committees on Govt. Reorganization, Flood Control, Water Resources and Control, and Committee on Corps of Engrs. Norman Medal from ASCE, 1902. Mem: ASCE (dir., 1908-10, vice pres., 1914-15), Am. Inst. of Consulting Engrs. (charter mem), representative of the Detroit Engr. Soc. on the Am. Engr. Council (vice pres., 1923-31), Am. Water Works Assn., New England Water Works Assn., Am. Assn. of Engrs., Western Assn. of Engrs., Michigan Engr. Soc., and the Detroit Bd. of Commerce, Univ. Clubs of Detroit and Ann Arbor. Publ: section on "Hydraulics" in the American Civil Engineer's Handbook, "Investigation of Transmission of Typhoid Fever in Detroit and St. Clair Rivers" (1897), Hydraulic Tables (1905). Refs: TASCE, WAB, Schnitter, WWE.

WILLIAMS, JESSE LYNCH; b. Westfield, NC, May 6, 1807; d. Ft. Wayne, IN, Oct. 9, 1886; f. Jesse; m. Sarah (Terrell); w. Susan Creighton. Moved with his family to Cincinnati, OH, 1814, Warren County then Wayne County, IN, 1819. Student at Lancasterian Seminary, Cincinnati. On 1st survey of Miami & Erie Canal in Ohio from Cincinnati to Maumee Bay, 1828. Made final location of canal from Locking Summit to Chillicothe and const. one division, including a dam and aqueduct across Scioto River. Mem. of bd. of engrs. who decided to use reservoirs rather than long feeders from district streams for supplying water to summit level of the canal. Chief engr. of Wabash & Erie Canal. In charge of all other canals in IN, 1835. Engr. in chief of all canal routes, 1836. In charge of RRs and turnpikes, 1836. Engaged in mercantile and manufacturing operations at Ft. Wayne, 1842-47. Chief engr. of Wabash & Erie Canal, 1847-76. Chief engr. of Ft. Wayne & Chicago RR, 1854-56. Dir. of Pittsburgh, Ft. Wayne & Chicago RR, 1856-73. Govt. dir. of Union Pacific RR, 1864-69. Report on const. and equipping road through Rocky Mountains led to the Credit Mobilier Investigation. Appt. receiver and engr. of the Grand Rapids & Indiana RR, Jan. 19, 1869-Oct. 1870. Asst. chief engr. in charge of completion of the Cincinnati, Richmond & Ft. Wayne RR, June 1871. Original dir. of Presbyterian Theological Seminary of the Northwest. Refs: DAB, WWW, CAB.

WILSON, JOHN ALLSTON; f. Phoenixville, PA, April 24, 1837; d. Jan. 19, 1896; f. William Hazell; m. Jane (Miller); w. Elizabeth H. Loyd; c. seven. Attended school in Philadelphia, 1842-49. Educated in private school in Chester County, PA, 1849-53. Entered RPI, Oct. 1853, grad. in July 1856 with degree of civ. engr. Served as topographer on surveys in Central America for the Honduras Interoceanic RR, April 1857-summer 1858. Asst. engr. on Pennsylvania RR, June 1858. Promoted to position of principal asst. engr. in charge of const., 1860-64. Chief engr. of Junction and Connecting RR Co., 1861-64. Aide on staff of Gen. D. N. Couch, cmdr. of Dept. of the Susquehanna, in charge of const. of fortifications at Harrisburg and vicinity, rebuilt Cumberland Valley RR, 1863. Appt. chief engr. for Pennsylvania RR Co., lessee of Philadelphia & Erie RR, March 1, 1864-68. Transferred to main line of Pennsylvania RR as chief engr. of maintenance of way, Jan. 1, 1868-70. Located and supt. of Low Grade Division of Allegheny Valley RR, 1870-75. Engr. of Bennett's Branch RR, April 1, 1874. Chief engr. for Pennsylvania RR in charge of const. of branch roads, const. Morrison's Cove Branches and extension of the Tyrone & Clearfield RR to Curriersville, PA, April 1870-April 1875. Consulting engr. in const. of centennial buildings in Philadelphia, 1875. Partner in Wilson Bros & Co., Philadelphia, Jan. 1, 1876-96. Chief engr. in North & West Branch RR of Staten Island Rapid Transit RR, of Buffalo Run RR, 1876. Manufactured lumber and mining bituminous coal in western PA, 1873. Designed and built Delaware Extension of the Pennsylvania RR including bridge over the Schuykill River at US arsenal in Philadelphia, Junction RR, Connecting RR, shops buildings and bridges on Philadelphia & Erie RR, the main line of Pennsylvania RR, Morrison's Cove and Clearwater extension branches of the Pennsylvania RR, Bennett's Branch North and West Branch RR, Buffalo Run RR. Mem: ASCE, Franklin Inst., Philadelphia Engrs. Club, Am. Inst. of Mining Engrs., Historical Soc. of PA, Philadelphia Art Club and the St. Andrew's Soc. Refs: RPI Bio, TASCE, CAB.

WILSON, JOSEPH MILLER; b. Phoenixville, PA, June 20, 1838; d. Philadelphia, Nov. 24, 1902; f. William Hasell; m. Jane (Miller); w. Sarah Pettit; c. Alice May, Mary Hasell. Studied in a family school under private tutors. Entered RPI, Sept. 1854, grad. with degree of civ. engr. in 1858. Studied analytical chemistry for 2 months with Prof. F. A. Genth in Philadelphia. Engr.

work on Pennsylvania RR, summer 1859, asst. engr., March 1860. Moved to Altoona, PA, 1860. Resident engr. on Middle Division, 1863-65. Principal engr., engr. of bridges and buildings over entire road in special charge of bridges, 1865-Jan. 1, 1886. Moved to Philadelphia, 1867. Engr. of bridges and buildings on Philadelphia, Wilmington & Baltimore RR. Won premium in competition for plans for Philadelphia Centennial Exhibition building, 1873. Made designs and took charge of erection of main exhibition buildings and machinery hall as engr. and architect. Organized firm of Wilson Bros. & Co., civ. engrs. and architects, June 1876. Engaged as expert for Brooklyn Bridge. Built buildings and bridges including the Susquehanna Bridge, Schuykill Bridge, Philadelphia Trenton Bridge, New Brunswick Bridge, Broad St. Station, Pennsylvania RR, Philadelphia Baltimore & Potomac Station, Washington, DC, Weehawken Station, NY, West Shore & Buffalo RR. M.A. from Univ. of Pennsylvania, 1867. Received medals and awards at Centennial Exhibition for plans of bridges and buildings, main exhibition building and machinery hall. Telford Premium from Inst. of Civ. Engrs. of Great Britain, 1878. Trustee and mem. of executive committee of the Pennsylvania Museum and School of Industrial Art. Chair of bd. of expert engrs. on Washington Aqueduct Tunnel and Reservoir, 1888-89. Chair of bd. of expert engrs. to report on RR terminal and station question at Providence, RI, 1888. Expert engr. examining and reporting on condition of elevated RRs in NYC. Expert engr. reporting to the Bd. of Rapid Transit Commissioners for NYC, 1891. Consulting engr. for Philadelphia & Reading RR CO. Reported in improvement of water supply for Philadelphia, 1899. Mem: ASCE (dir., 1888-89, vice pres., 1894-96), Soc. of Civ. Engrs. of London, Am. Inst. of Architects, Franklin Inst. (mgr. 1868-87, pres., 1887-97), Philadelphia Engrs. Club (pres., 1888), Am. Assn. for Advancement of Science, Century Assn. of NYC, Am. Philosophical Soc., Photographical Soc. of Philadelphia. Publ: "Mechanical and Scientific" and "Historical" sections of the illustrated catalogue for the Centennial Exhibition of 1876, "Bridge Over the Monongahela River at Port Perry, PA" (1880), "American Permanent Way" (1885), "On Specification for Strength of Iron Bridges" (1886), "On Schools With Particular Reference to Trade Schools" (1890), "The Philadelphia and Reading Terminal RR and Station in Philadelphia" (1895), and several papers for scientific and engr. journals. Refs: RPI Bio, TASCE, WAB, WWW, DAB, CAB, CCI.

WOLMAN, ABEL; b. Baltimore. MD, June 1892; d. Baltimore, MD, Feb. 1989; f. Morris; m. Rosa; c. Markley Gordon. Grad. The Johns Hopkins Univ., B.A., 1913; B.S. 1915. Worked as asst. div. engr., MD State Dept. of Health, 1915-22; chief engr., 1922-39. Chair, MD State Planning Commission, 1934-45. Chair, Water Resources Committee, National Resources Planning Bd., 1935-41. Prof. and chair, Dept. of Sanitary Engr., The Johns Hopkins Univ. School of Engr. and School of Hygiene and Public Health, 1937-62; Prof. Emeritus, 1962-89. Consultant, US Public Health Service, 1939-66. Consulting engr., Tennessee Valley Authority, 1939-76. Water supply consultant, Bethlehem Steel, Sparrows Point, MD, 1940-64. Mem., National Technical Advisory Committee of the Secy. of War, 1941-49. Consultant, National Resources Planning Bd., 1942-45. Chief consultant to Dir. of War Utilities, War Production Bd., 1943-45. Consultant, Bd. of Sanitary Control and Protection of Public Water Supply for NYC, 1950, 1965-66; chair, 1951-53. Advisor, US Geological Survey, 1943-67. Chair, Expert Committee on Environmental Sanitation of the World Health

Organization, 1946-50. Mem., Advisory Committee on Reactor Safeguards, 1947-50, Atomic Safety and Licensing Bd. Panel, 1960-72, US Atomic Energy Commission, 1947-50. Consultant, govt. of Ceylon, 1955; chair, Water Supply Mission to Taiwan, 1961; consultant on Drinking Water and Sanitation Decade, 1979-88 for the World Health Organization. US delegate, Geneva Conference on Peaceful Uses of Atomic Energy, 1958- . Chair, Bd. of Consultants for Water Supply, Calcutta, India, 1959- . Consultant, Metropolitan Sao Paulo, Brazil for US A.I.D., 1963-68. Mem., National Panel of Arbitrators, Am. Arbitration Assn., 1968-72. Consultant, National Water Commission, Washington, DC, 1969-73. Consultant, New Jersey Master Water Plan, 1975-80. Chair, Bd. of Consultants on Safety of Dams, Miami Conservancy District, 1976-77. Chair, Bd. of Review, Diversion of Mediterranean to the Dead Sea, State of Israel, 1981- . Awards: 5 hon. doctorates, including The Johns Hopkins Univ., 1937 and the Drexel Inst., 1957; Lasker Award of the Am. Public Health Assn., 1960; National Medal of Science, 1975; Tyler Ecology Award, 1976; Environmental Regeneration Award of the Dubos Center for Human Environments, 1975; Health for All Medal of the World Health Organization, 1988; Ben Gurion Award of the State of Israel, 1976; Horton Medal of the Am. Geophysical Union, 1986; John Wesley Powell Award, US Geological Survey, 1986; fellow, National Academy of Engr. and National Academy of Science. Mem: Am. Public Health Assn. (chair, executive bd., 1939-49, pres., 1939), Am. Water Works Assn. (bd. of dir., 1939-43, pres., 1942). Editor-in-chief, *Journal of the American Water Works Association*, 1921-37 and *Municipal Sanitation*, 1929-35. Editor, *American Journal of Public Health*, 1955. Publ: 31 of his more than 200 publications are collected in *Water, Health snd Society: Selected Papers*, edited by Gilbert White (Bloomington, IN: Indiana Univ. Press, 1969).

WOOD, DEVOLSON; b. Smyrna, NY, June 1, 1832; d. Hoboken, NJ, June 27, 1897; f. Julius; m. Amanda (Billings); w. Cordera E. Crawe; w. 2nd Frances Hartson; c. seven. Taught in a public school. Studied at Cazenovia and Albany, NY and at the RPI, grad. 1857 with degree of civ. engr. Prof. of civ. engr. at Univ. of Michigan, 1857-72. Chair of mathematics and mechanics in Stevens Inst., Hoboken, NJ, 1872-85. Transferred to chair of engr., 1885-97. Designed ore dock at Marquette, MI, 1866. Invented an air compressor and a steam rock drill. Pres. of Bd. of Education in Boonton, NY, 1881. Hon. A.M. from Hamilton College, 1859, and M.S. from Univ. of Michigan, 1859. Mem: ASCE, Soc. of Mech. Engrs., Am. Assn. for the Advancement of Science, Soc. of Architects. Publ: *Treatise on the Resistance of Materials* (1871), *A Treatise on the Theory of the Construction of Bridges and Roofs* (1872), *The Elements of Analytical Mechanics* (1876), *Principals of Elementary Mechanics* (1878), *The Elements of Co-ordinate Geometry* (1879), *The Mechanics of Fluids* (1884), *Trigonometry, Analytical Plane and Spherical* (1885), *Thermodynamics* (1887). Refs: WAB, CAB, WWW, RPI Bio.

WORCHESTER, JOSEPH RUGGLES; b. Waltham, MA, May 9, 1860; d. May 9, 1913; f. Benjamin; m. Mary Clapp (Ruggles); w. Alice Jeanette Wheeler; c. Alice Martha, Barbara, Thomas, Ruth Hunt. Grad. from Harvard Univ., 1882 with degree of A.B. Engr. dept. and chief engr., Boston Bridge Works, 1882-94. In private practice, 1894-1906. Member of the firm of J. R. Worcester & Co., 1906-24. Consultant for Boston Transit Commission in subway const., Boston Terminal Co. in const. of train shed of South Station, Boston Elevated RR elevated structure among others. Designed bridge over Connecticut River at Bellows Falls, VT, and bridge between Portland and South Portland, ME. Pres. of trustees of the Waltham Hospital, Trustee of the Waltham Savings Bank. Mem: ASCE, Am. Academy of Arts and Sciences, Boston Soc. of Civ. Engrs., Am. Soc. for Testing Materials, Am. Railway Engr. Assn., Am. Inst. of Consulting Engrs., Am. Concrete Inst., Boston Chamber of Commerce, New Church Inst. of Education. Refs: WWE, WWW.

INDEX OF BIOGRAPHIES

NOTE: An asterisk indicates biography is in Volume I

*Abbot, Henry Larcom
*Abert, John James
Abrams, Duff Andrew
*Adams, Julius Walker
Alden, John Ferris
Allen, Calvin Francis
*Allen, Horatio
Allen, Kenneth
Ammann, Othmar Hermann
Anderson, John Francis
*Bache, Alexander Dallas
*Bache, Hartman
*Bacon, Nathaniel Terry
Baker, George Titus
*Baker, Ira Osborn
Baker, William Edgar
*Baldwin, Loammi, Sr.
*Baldwin, Loammi, Jr.
*Banneker, Benjamin
Barnard, John Fiske
Barnard, John Gross
Barnes, Oliver Weldon
Bassett, Carrol Phillips
*Bates, David Stanhope
Bates, Onward
Bayles, James Copper
Beahan, Willard
Beardsley, James Wallace

Benham, Henry W.
Bensel, John Anderson
Benzenberg, George H.
Berg, Walter Gilman
*Bernard, Simon
Bigelow, Charles H.
Black, William Murray
*Bogardus, James
Bogart, John
*Bogue, Virgil Gay
*Boller, Alfred Pancoast
*Bollman, Wendel
Bond, Edward Austin
*Bonzano, Adolphus
Borden, Simeon
*Bouscaren, Louis Gustave
*Boyden, Uriah Atherton
Brinckerhoff, Henry M.
*Bro(a)dhead, Charles C.
Broadhead, Garland Carr
Brown, Charles Carroll
Brown, William Henry
*Brunel, Marc Isambard
*Bryant, Gridley
*Buck, Leffert Lefferts
*Buckout, Isaac Craig
*Burden, Henry
*Burr, Theodore
*Burr, William Hubert
*Burr, William Austin
*Burt, William Austin
Bush, Lincoln
Cain, William
Campbell, Allan

Carll, John Franklin
Casey, Thomas Lincoln
Cass, George Washington
Cassatt, Alexander J.
*Chanute, Octave
*Chesbrough, Ellis S.
*Childe, John
Chittenden, Hiram Martin
Church, George Earl
Clyde, George Dewey
Cogswell, William Browne
*Cohen, Mendes
*Colles, Christopher
Collingwood, Francis
Cooley, Lyman Edgar
*Cooper, Theodore
*Corthell, Elmer L.
*Cox, Lemuel
Crandall, Charles Lee
*Craven, Alfred Wingate
*Croes, John James R.
*Crozet, Claude
Curtis, Samuel Ryan
*De Brahm, William G.
DeLacy, Walter W.
Delafield, Richard
Detmold, Christian E.
Devereaux, John Henry
*De Witt, Simeon
Dillon, Sidney
*Doane, Thomas
*Dodge, Grenville Mellen
Donovan, John Joseph
Douglas, Walter Jules

Douglass, David Bates
Duane, James Chatham
*Duportail, Chevalier L.
Durfee, William Franklin
Durham, Caleb Wheeler
*Eads, James Buchanan
*Eaton, Amos
*Ellet, Charles, Jr.
*Ellicott, Andrew
Endicott, Mordecai T.
Ernst, Oswald Herbert
Estabrook, John Davis
Evans, Anthony Walton
*Evans, Oliver
*Fanning, John Thomas
*Farnum, Henry
Felton, Samuel Morse Sr.
Felton, Samuel Morse Jr.
*Ferris, George W. G.
*Fink, Albert
Finley, James
*FitzGerald, Desmond
*Flad, Henry
*Francis, James Bicheno
Freeman, John Ripley
Frizell, Joseph Palmer
*Fry, Joshua
*Fteley, Alphonse
*Fuertes, Estevan A.
*Fulton, Robert
Gagel, Edward
Gaillard, David DuBose
*Geddes, James
*Gerstner,Franz A R.von

Giaver, Joachim G.
*Gillespie, William M.
*Gillmore, Quincy Adams
*Goethals, George W.
Gotshall, William Charles
Graff, Frederic
*Graff, Frederic(k)
Graham, Charles Kinnaird
Grant, Edward Maxwell
*Gratoit, Charles
Greene, Charles Ezra
Greene, David Maxson
*Greene, George Sears
Greene, George Sears
Greiner, John Edwin
*Gridley, Richard
Grinnell, Frederick
Grow, Henry
Grunsky, Carl Ewald
*Gzowski, Casimer S.
*Halladie, Andrew Smith
*Hammond, John Hays
Harris, Daniel Lester
Harrod, Benjamin Morgan
*Hassler, Ferdinand R.
Haswell, Charles Haynes
*Haupt, Herman
Haupt, Lewis Muhlenberg
*Haviland, John
Hebert, Paul Octave
Henck, John Benjamin
Henny, David Christian
*Hering, Rudolph
*Hermany, Charles

*Herschel, Clemens
Hildenbrand, Wilhelm
*Hilgard, Julius Erasmus
Hoadley, John Chipman
Hodges, Harry Foote
Hogeland, Albert Harrison
*Holley, Alexander L.
*Holme, Thomas
Holmes, Howard Carleton
Houghton, James Franklin
*Howe, William
Humphreys, Andrew A.
Hunt, Robert Woolston
*Hutchins, Thomas
*Hutton, William Rich
*Hyatt, Thaddeus
Ingersoll, Colin Macrae
*Isherwood, Benjamin F.
Jacobs, Charles Matthias
*Jacoby, Henry S.
*Jenney, William LeBaron
*Jervis, John Bloomfield
Johnson, Edwin Ferry
Johnson, John Butler
Jorgenson, Lars R.
*Judah, Theodore Dehone
Judson, William Pierson
*Katte, Walter,
*Kay, Edgar Boyd
*Keefer, Samuel
Kimball, George Henry
Kingman, Lewis
Kingsley, William
*Kinnicutt, Leonard P.

*Kirkwood, James Pugh
Kittredge, George Watson
*Kneass, Samuel H.
Kneass, Strickland
*Knight, Jonathan
Knowles, Morris
*Kosciusco, Thaddeus
Koyl, Charles Herschel
Landreth, Olin Henry
*Latrobe, Benjamin H. Sr.
*Latrobe, Benjamin H. Jr.
Latrobe, Charles H.
Laurie, James
*L'Enfant, Pierre Charles
Lewis, William Gaston
*Lindenthal, Gustav
*Linville, Jacob Hays
*Long, Stephen Harriman
Lovell, Mansfield
Ludlow, William
Lundie, John
McCallum, Daniel Craig
MacDonald, Charles
MacDonald, Thomas H.
*McAlpine, William J.
McClellan, Carswell
McMath, Robert Emmet
McMillan, Charles
*McNeill, William Gibbs
*Mahan, Dennis Hart
Mahone, William
Main, Charles Thomas
*Mansfield, Jared
Marshall, William Louis

Mason, Clairbourne Rice
Mason, William Pitt
Maynard, George William
*Mead, Elwood
Meade, George Gordon
*Meiggs, Henry
*Meigs, Montgomery C.
*Menocal, Aniceto Garcia
Merrill, William Emery
*Merriman, Mansfield
Miller, Ezra
Millington, John
*Mills, Hiram Francis
Milner, John Turner
Moore, Robert
Morell, George Webb
*Morison, George S.
Morris, Thomas A.
Morse, Charles Adelbert
Morse, Edwin Kirtland
*Mulholland, William
Murphy, John Wilson
Nettleton, Edwin S.
Newton, John
Nichols, Othniel Foster
*Noble, Alfred
Northen, William Ezra
Nostrand, Peter Elbert
Ockerson, John Augustus
Olcott, Eben Erskine
Osborn, Frank Chittenden
Page, Logan Waller
Palfrey, John Carver
*Palmer, Timothy

Pardee, Ario
*Partridge, Alden
Parsons, William Barclay
*Pegram, George H.
Peters, Richard
*Pettit, Henry
Plympton, George W.
Poe, Orlando Metcalfe
*Post, Simeon
Pratt, Thomas Willis
Purdon, Charles de la
    Cherois
Purdy, Corydon Tyler
Rafter, George W.
Randolph, Isham
Ranney, Henry Joseph
Ransome, Ernest L.
*Rea, Samuel
Renwick, Henry B.
*Renwick, James
Rice, George Staples
Ricketts, Palmer C.
*Rittenhouse, David
*Roberdeau, Isaac
Roberts, Benjamin Stone
Roberts, Nathan Smith
*Roberts, Solomon White
*Roberts, William Milnor
Robinson, Albert Alonzo
*Robinson, Moncure
Robinson, Stillman W.
Rodd, Thomas
*Roebling, John A.
*Roebling, Washington A.

*Romans, Bernard
Rosewater, Andrew
Rotch, William
*Rowland, Thomas Fitch
St. John, Isaac Monroe
Sayre, Robert Heysham
Schneider, Charles C.
Schuyler, James Dix
Searles, William Henry
Serrell, Edward Wellman
Shankland, Edward Clapp
*Shedd, Joel Herbert
Shunk, William Findlay
Sibert, William Luther
Sickles, Frederick E.
Sidell, William Henry
Simpson, James Hervey
Slifer, Hiram Joseph
Sloan, Robert Imlay
Smith, Charles Shaler
Smith, Gustavus Woodson
Smith, Hamilton
*Smith, William Sooy
*Sprague, Frank Julian
*Staley, Cady
*Stevens, John
*Stevens, John Frank
*Stone, Amasa
*Storrow, Charles S.
*Strickland, William
*Sullivan, John Langdon
*Sutro, Adolph Heinrich
Swain, George Fillmore
*Swift, Joseph Gardner

Talbot, Arthur Newall
*Talcott, Andrew
Tatham, William
Terzaghi, Karl
*Thacher, Edwin
*Thayer, Sylvanus
*Thomson, John Edgar
Thompson, Sanford E.
Thomson, Reginald Heber
Totten, George Muirson
*Totten, Joseph Gilbert
*Town, Ithiel
*Trautwine, John C., Sr.
Trautwine, John C., Jr
Trimble, Issac Ridgeway
Tsechebotareff, Gregory
*Turnbull, William
Turner, Claude Allen P.
Tweeddale, William
*Viele, Egbert Ludovicus
Vose, George Leonard
Waddell, John Alexander
Waite, Christopher C.
Walker, Reuben Lindsay
*Wallace, John Findley
*Waring, George Edwin
Warren, Gouveneur K.
Watkins, John Elfreth
Webb, George Herbert
Webster, George Smedley
*Wegmann, Edward
Weitzel, Godfrey
*Welsh, Ashbel
*Wellington, Arthur M.

*Wernwag, Lewis
Weston, Charles Valentine
*Weston, William
*Whipple, Squire
*Whistler, George W.
*White, Canvass
*Whitney, Asa
Whitney, Charles Smith
*Whittemore, Don Juan
Williams, Gardner S.
Williams, Jesse Lynch
*Williams, Jonathan
Wilson, John Allston
Wilson, Joseph Miller
*Wilson, William Hasell
*Winans, Ross
Wood, DeVolson
Worchester, Joseph R.
*Worthen, William Ezra
*Wright, Benjamin
*Wright, Benjamin Hall